110
Structure and Bonding

Series Editor:
D.M.P. Mingos

Springer

Berlin
Heidelberg
New York
Hong Kong
London
Milan
Paris
Tokyo

Applications of Evolutionary Computation in Chemistry

Volume Editor:
R. L. Johnston

With contributions by
H.M. Cartwright · V.J. Gillet · S. Habershon
K.D.M. Harris · B. Hartke · R.L. Johnston · R. Unger
S.M. Woodley

Springer

The series *Structure and Bonding* publishes critical reviews on topics of research concerned with chemical structure and bonding. The scope of the series spans the entire Periodic Table. It focuses attention on new and developing areas of modern structural and theoretical chemistry such as nanostructures, molecular electronics, designed molecular solids, surfaces, metal clusters and supramolecular structures. Physical and spectroscopic techniques used to determine, examine and model structures fall within the purview of Structure and Bonding to the extent that the focus is on the scientific results obtained and not on specialist information concerning the techniques themselves. Issues associated with the development of bonding models and generalizations that illuminate the reactivity pathways and rates of chemical processes are also relevant.

As a rule, contributions are specially commissioned. The editors and publishers will, however, always be pleased to receive suggestions and supplementary information. Papers are accepted for *Structure and Bonding* in English.

In references *Structure and Bonding* is abbreviated *Struct Bond* and is cited as a journal.

Springer WWW home page: http://www.springeronline.com
Visit the SB home page at http://www.springerlink.com

ISSN 0081-5993
ISBN 3-540-40258-6
DOI 10.1007/b10607
Springer-Verlag Berlin Heidelberg New York

CIP Data applied for

Springer-Verlag is a part of Springer Science+Business Media

springeronline.com

Typesetting: Fotosatz-Service Köhler GmbH, Würzburg
Production editor: Christiane Messerschmidt, Rheinau
Cover: Künkel+Lopka, Heidelberg/design & production GmbH, Heidelberg

Printed on acid-free paper 02/3020 – 5 4 3 2 1 0

Series Editor

Professor D. Michael P. Mingos
Principal
St. Edmund Hall
Oxford OX1 4AR, UK
E-mail: michael.mingos@st-edmund-hall.oxford.ac.uk

Volume Editor

Dr. Roy L. Johnston
School of Chemistry
University of Birmingham
Edgbaston
Birmingham B15 2TT, UK
E-mail: roy@tc.bham.ac.uk

Editorial Board

Structure and Bonding is also Available Electronically

For all customers who have a subscription to Structure and Bonding, we offer the electronic version via SpringerLink free of charge. Please contact your librarian who can receive a password for free access to the full articles by registration at:

http://www.springerlink.com

If you do not have a subscription, you can still view the tables of contents of the volumes and the abstracts of each article by going to the SpringerLink Homepage, clicking on "Browse by Online Libraries", then "Chemical Sciences", and finally choose Structure and Bonding.

You will find information about the

- Editorial Board
- Aims and Scope
- Instructions for Authors
- Sample Contribution

at http://www.springeronline.com using the search function.

Preface

The field of Evolutionary Computation covers a wide range of computational approaches, including: Evolutionary Algorithms – of which Genetic Algorithms are probably the most widely studied; Evolutionary Strategies; and Evolutionary or Genetic Programming. What these approaches have in common is that they are based (more or less closely) on principles of biological evolution – using operators that are analogues of natural processes (such as genetic crossover, mutation and natural selection) to explore multi-dimensional parameter spaces and perform multi-variable optimisation [1–3]. In the case of Evolutionary/Genetic Programming, computer programs, algorithms or functions are evolved from simpler sub-programs, mathematical operations or functions [3].

Evolutionary Computation has developed over the past 30 years or so, with initial applications being primarily in computer science and in modelling adaptive processes in natural systems. Since the early 1990s, however, Evolutionary Computation – in particular Genetic Algorithms – have been increasingly used in a variety of global optimisation problems in chemistry, as well as in physics, materials science and biology [4–6].

In this Volume, Evolutionary Computation methods are introduced and their application to a number of areas of current chemical interest is reviewed. Firstly, the differences between the various methods of Evolutionary Computation and the principles underlying them are outlined in the Chapter by Hugh Cartwright, which also discusses, in a generic way, the pseudo-evolutionary operators used in each case. Specific applications and detailed discussions of algorithms and methodologies are presented in subsequent Chapters.

Nanoscience is an important modern research area, which spans chemistry, physics, materials science and biology. Evolutionary Algorithms have found a number of applications in nanoscience, the most popular being in the geometry optimisation of atomic and molecular clusters and nanoparticles, as discussed in the Chapter by Bernd Hartke. In the area of chemical crystallography, there has been considerable success in applying Evolutionary Computation techniques to the determination of structures from single crystal, powder and thin film diffraction data – with applications to organic, inorganic and even biological molecules. Some of this work is reviewed in the Chapter by Kenneth Harris, Scott Habershon and myself. The related problem of crystal structure prediction (which may additionally be used to assist in structure determination) is also of great interest and is the subject of the Chapter by Scott Woodley.

An increasingly active area of application of Evolutionary Computation is in bioinformatics and biological chemistry, representing a "return" of artificial evolutionary methods to the study of real natural systems. Evolutionary Algorithms have been widely used in the field of drug design – for example in combinatorial library design and the derivation of quantitative structure-activity relationship (QSAR) models, as described in the Chapter by Valerie Gillet. The application of Genetic Algorithms to the prediction of the three-dimensional secondary and tertiary structures of proteins (from their one-dimensional sequences alone) – i.e. attempting to solve the so-called "protein folding problem" – is also a fertile area of Evolutionary Computation and is reviewed in the Chapter by Ron Unger.

While we have attempted to present an overview of some important applications of Evolutionary Computation in chemistry (and biochemistry), this is not meant to be a comprehensive treatment. Other significant applications of Evolutionary Algorithms in chemistry include: chemometrics (the use of mathematics and computer science to control and optimise chemical processes, to analyse and mine chemical data, and to investigate chemical systems); deconvolution and interpretation of spectroscopic data; optimisation of laser parameters and shaping of laser pulses for optimal "quantum control" of chemical reactions; parameterisation and optimisation of potential energy functions; solution of the quantum mechanical Schrödinger equation; structural studies of RNA and DNA; pharmacophore mapping and receptor modelling. A particularly useful source of references on the application of Evolutionary Computation to a wide range of chemical problems can be found in the bibliographic database maintained by David Clark [7].

It is very likely that the future will see an expansion in the application of pseudo-evolutionary techniques for optimisation problems of increasing complexity, in chemistry and other sciences, becoming widely used tools in computational chemistry – rather like Molecular Dynamics and Monte Carlo simulation methods have become. In addition to Evolutionary Computation, recent years have witnessed an explosion in computational algorithms and methodologies based on Artificial Intelligence (e.g. Artificial Neural Networks, Expert Systems and Fuzzy or Soft Computing) [6] and other techniques derived by analogy with natural systems (e.g. Artificial Immune Systems [8] and Swarm Intelligence – as in Ant Colony Optimization [9]). Some of these techniques have already been applied to chemical problems [4–6] and many more applications are likely in the future.

Birmingham, November 2003 R. L. Johnston

1. Holland J (1975) Adaptation in natural and artificial systems, University of Michigan Press, Ann Arbor, MI
2. Goldberg DE (1989) Genetic algorithms in search, optimization and machine learning, Addison-Wesley, Reading, MA
3. Koza JR (1992) Genetic programming: on the programming of computers by means of natural selection, MIT Press, Cambridge, MA

4. Cartwright HM (1993) Applications of artificial intelligence in chemistry, Oxford University Press, Oxford
5. Clark DE (ed.) (2000) Evolutionary algorithms in molecular design, Wiley-VCH, Weinheim
6. Cartwright HM, Sztandera LM (eds.) (2003) Soft computing approaches in chemistry, Springer-Verlag, Berlin
7. Clark DE (2003) Evolutionary algorithms in computer-aided molecular design, http://panizzi.shef.ac.uk/cisrg/links/ea_bib.html
8. De Castro LN, Timmis, J (2002) Artificial immune systems: a new computational intelligence approach, Springer-Verlag, Berlin
9. Bonabeau E, Dorigo M, Theraulaz G (1999) Swarm intelligence: from natural to artificial systems, Santa Fe Institute Studies on the Sciences of Complexity, Oxford University Press, New York

Contents

An Introduction to Evolutionary Computation and Evolutionary Algorithms
H. M. Cartwright 1

Application of Evolutionary Algorithms to Global Cluster Geometry Optimization
B. Hartke 33

Applications of Evolutionary Computation in Structure Determination from Diffraction Data
K. D. M. Harris, R. L. Johnston, S. Habershon 55

Prediction of Crystal Structures Using Evolutionary Algorithms and Related Techniques
S. M. Woodley 95

Applications of Evolutionary Computation in Drug Design
V. J. Gillet 133

The Genetic Algorithm Approach to Protein Structure Prediction
R. Unger 153

Author Index 101–110 177

Subject Index 181

Structure and Bonding, Vol. 110 (2004): 1–32
DOI 10.1007/b13931

An Introduction to Evolutionary Computation and Evolutionary Algorithms

Hugh M. Cartwright

Physical and Theoretical Chemistry Laboratory, Oxford University, South Parks Road, Oxford OX1 3QZ, UK
E-mail: hugh.cartwright@chem.ox.ac.uk

Abstract Numerous techniques exist in science for tackling large-scale optimisation problems. In many instances, scientists must identify the best solution from within a pool containing millions of possible solutions. Evolutionary algorithms are particularly adept at handling this sort of optimisation and their advantages often become more marked as the size of the search space grows. This chapter introduces evolutionary methods, and outlines some of the principles of genetic algorithms, genetic programming and evolution strategies.

Keywords Genetic algorithm · Genetic programming · Evolutionary algorithm · Evolution strategies · Optimisation

1 Introduction 2

1.1 Problem-Solving 2
1.2 Deterministic and Stochastic Methods 3

2 Search 4

2.1 Optimisation 4
2.2 The Role of Learning in Problem-Solving 6
2.3 Dumb and Slightly Less Dumb Searches 6

3 Evolutionary Algorithms 11

3.1 Change and Competition 11
3.2 A Simple Evolutionary Algorithm 12
3.2.1 Initial Population 12
3.2.2 Selection 14
3.2.3 Swapping of Genetic Material 14
3.2.4 Mutation 15
3.3 Population Drift 16
3.4 Evolutionary Variables 17

4 Genetic Algorithms 18

4.1 Strings – the Genetic Algorithm Solution 18
4.2 Genetic Algorithm Operators 20
4.2.1 Initial Population 20
4.2.2 Fitness 21

4.2.3 Selection . . . 21
4.2.4 Crossover . . . 23
4.2.5 Mutation . . . 24

5 **Evolution Strategies** . . . 25

5.1 Basic Evolution Strategies . . . 25
5.2 Population-Based ES . . . 26

6 **Genetic Programming** . . . 28

6.1 Hands-Off Programming . . . 28
6.2 Genetic Programming Individuals . . . 28
6.3 Genetic Programming Operators . . . 29

7 **Conclusion** . . . 31

8 **Further Reading** . . . 32

Abbreviations

AI Artificial intelligence
EA Evolutionary algorithm
ES Evolution strategies
GA Genetic algorithm
GP Genetic programming

1 Introduction

1.1 Problem-Solving

Problem-solving is the most fundamental of scientific tasks. At its heart is the search for an understanding of the behaviour of the physical world, which, in its most comprehensive form, brings together both descriptive and quantitative elements. It is the search for solutions to scientific problems which is the central theme of this volume. In the chapters that follow, chemists discuss recent progress in the application of some of the most promising algorithms now available to solve scientific problems.

A quite remarkable diversity of computational challenges exists in science. They range from the calculation of the strains in engineering structures or the interpretation of the NMR spectra of biological samples, to modelling the evolution of star clusters or predicting the way in which pollution moves though groundwater courses. It would simplify the work of scientists if, when a new scientific problem was encountered, the choice of an appropriate method of solution was clear-cut, determined just by the problem type; however, this is rarely the case. Some of the trickiest problems in chemistry, such as the prediction of the three-dimensional in vivo structure of a protein from its amino acid sequence, are so demanding that they are not yet handled entirely satisfactorily by any method.

Within the rich set of scientific problems are many that are difficult to solve. It is clearly unrealistic to expect that any single method can invariably be successful, no matter what the type of problem to which it is applied, so scientists must be pragmatic, taking advantage of whatever algorithmic tools are available and seem likely to be productive.

A broad range of methods, developed and refined through repeated use, exist to tackle problems in chemistry. Most are well understood; their limitations and strengths have become apparent through extensive investigation and the circumstances under which they can be applied effectively are well documented. Many are deterministic, performing in a manner which is reproducible from one invocation to the next, so if they are applied repeatedly to the same problem identical results are obtained. These methods therefore provide unambiguous (though not necessarily correct) solutions to the problems to which they are applied, and some allow the estimation of a probable error.

Analytical methods may deliver the best solution of which they are capable in a single step. Typical of a simple analytical approach is the calculation of the best least-squares straight line through a two-dimensional set of data points. The standard formulae are used, and no refinement of the gradient and intercept is then required (or possible). By contrast, many nonanalytical methods, including such elementary techniques as SIMPLEX or hill-climbing, are iterative and may be of value when no analytical method is available. A range of quantum mechanical calculations, some of them amongst the most challenging and important in chemistry, are essentially iterative in nature because of the lack of any exact analytical method.

It is a characteristic of most iterative methods that their performance is strongly correlated with the values chosen for adjustable parameters. Thus, the step size in a hill-climb must be selected with care (and perhaps adjusted as the calculation proceeds) to ensure that movement across the search landscape is neither so languid that the task of locating the maximum is unreasonably drawn out, nor so volatile and unpredictable that the search is unable to settle on a maximum. Mathematical recipes sometimes exist which specify how the values of adjustable parameters in iterative algorithms should be chosen, or how they can be optimised as the calculation proceeds.

1.2 Deterministic and Stochastic Methods

It is not uncommon for an algorithm to be able to guarantee that it will locate the optimum solution, provided that it is set to work on an appropriate problem, and this is an important benefit. However, the established methods of solution are not without their drawbacks. Even when a guarantee does exist that the optimum solution can be found, it may not be possible to find this solution within an acceptable period of time. Deterministic methods are valuable if the solution which they provide is of sufficient quality, but they will, of course, fail in a reproducible fashion if they are unable to locate a sufficiently good solution on their first attempt.

As we shall see, this volume concentrates on methods which are stochastic, rather than deterministic, in nature – that is, the route that they take in seeking a solution depends in some way upon chance (and, as an important consequence, the solution found may vary from one invocation of the algorithm to the next). Although stochastic methods are not as familiar to most chemists as analytical methods, examples of their use are still widespread. Monte Carlo calculations, which combine a statistical approach with the chance nature of random events, have been widely used in fields such as molecular dynamics, but Monte Carlo is only suited to a restricted range of problems.

It is evident that no way of locating solutions is perfect, or universally applicable. Although established methods are often valuable, opportunities exist for newly developed techniques to outperform them, especially in the solution of some of the more challenging, and perhaps more interesting, problems in science. A number of approaches within the field of artificial intelligence (AI) show great promise in science, and scientific applications of AI are already numerous. Applications from within one of its principle areas, that of evolutionary algorithms (EAs), form the topic of this volume.

2 Search

2.1 Optimisation

Few companies are as dependent upon research as those in the pharmaceutical industry. The development of a new drug is enormously costly, so companies invest heavily in computational approaches to drug design, with the aim of reducing the number of chemicals that need to be screened for therapeutic activity, and thereby shortening the time to market for new drugs. Typically, pharmaceutical companies use structure-based design to address computational questions such as

- What compounds of molecular weight below 250 Da will fit into the active site of protein Y and will bind strongly to it?

The number of compounds whose molecular weight is less than 250 Da is, of course, vast, and many of these compounds will be capable of fitting into the active site of a given protein and binding to it. However, these potentially valuable molecules will be hugely outnumbered by those which cannot fit into the active site, or which bind to it only weakly. To identify compounds of potential therapeutic interest, the relatively tiny proportion of molecules which meet the specified criteria need to be pinpointed among the huge number of unsuitable molecules.

We can picture the algorithm whose task is to pick out the potentially valuable molecules as wandering across an extensive search space, on which are located somewhere all possible molecules. The best molecule, or optimum solution in this space, is sought, so the problem of identifying that molecule is one of optimisation – a term which is used frequently in evolutionary computing.

The search space in a problem of this sort is much too large to permit an exhaustive search, in which every solution is inspected individually. Problems which have similarly large search spaces abound in chemistry, and if the search across any very large space is to succeed, the algorithm must be able to locate optima even though it is able to sample just a small fraction of that space. It can only do this if it is able somehow to identify the best places to look, that is, promising regions of the search space in which it is likely that the best solutions are located.

A variety of tools exist to tackle optimisation problems. A common approach in such disparate fields as the social sciences and analytical chemistry is through the use of heuristics (rule-based approaches). This leads to an increasingly focused search, in which the region over which searching takes place is narrowed as successively more fine-grained rules are applied. Heuristic searches may be only modestly successful if the search space is large, since at an early stage they may limit the region over which they search to a small fraction of the whole. This can lead to the premature rejection of potentially valuable solutions. The ability at all stages in the calculation to inspect regions of the search space which are widely separated is a property of most forms of EAs, and is one of their key features.

This volume discusses some iterative optimisation methods drawn from within AI. Iterative methods of the sort discussed here are often computationally intensive, and as long as computers were of modest power these algorithms struggled to compete with other methods. Indeed, their computational demands had the effect of severely limiting the interests of scientists in them until the last decade of the twentieth century. However, evolutionary methods have, as we shall see, special advantages and they have become increasingly attractive as computer power has grown.

Table 1 Typical uses of artificial intelligence in chemistry. Genetic algorithm (*GA*), neural network (*NN*), self-organising map (*SOM*), knowledge-based system (*KBS*)

Problem	Artificial intelligence method
Chemical flow-shop control	GA
Molecular structure optimisation	GA
Intelligent sensors, electronic noses	NN
Analytical methods advisor	KBS
Ion chromatography	NN+GA
Polymer design	GA
Spectral deconvolution	GA
Spectral recognition	NN, SOM
Combinatorial catalysis	NN
Ceramic design	NN
Quantitative structure – activity relationship	GA
De Novo design	GA
Chirality – structure linkage	SOM
Waste minimisation	KBS

The boundaries to AI, within which evolutionary optimisers lie, are woolly, and this may make it appear a mysterious subject to those who are unfamiliar with it; even those whose research lies in AI find it hard to agree on the precise boundaries to the field. It encompasses a large family of algorithms, from neural networks and knowledge-based systems to those within evolutionary computing. Almost every major AI technique is now used within chemistry (Table 1), and in an increasing number of cases, AI is the method of choice.

2.2 The Role of Learning in Problem-Solving

At the core of AI is machine learning. It may seem strange that scientists use programs that learn as a tool to solve chemical problems; machine learning is more often associated with the development of intelligent robots than with the solution of problems in the physical sciences. However its methods have potential far beyond robotics.

Programs based on machine learning are designed in such a way that they can modify their behaviour as they run. This permits them to take advantage of their previous experience so that, as the way in which solutions are laid out across the search space becomes apparent, their strategy can be adjusted. Such programs might seem eccentric by comparison with conventional algorithms, and it is not immediately obvious why a computer program which needs to learn about a scientific problem before solving it might be more effective than one which relies upon familiar tools such as calculus. However, methods in machine learning have impressive power, and it is therefore understandable that they should attract the interest of scientists.

Just how does learning help in the solution of optimisation problems? Let us consider an example which, while it is conceptually a very simple task, does indicate why knowing a bit about how a search has gone in the past can help us to make better decisions in the future.

2.3 Dumb and Slightly Less Dumb Searches

The shade of dyes can be specified using a colour index, which defines the proportion of three component colours required to make any other colour. (This is analogous to the designation with which computer users are familiar with colours in terms of red/green/blue values.) If the component colours are placed on the circumference of a circle, other colours can be represented as points on or within that circle – a so-called colour wheel (Fig. 1).

Suppose that a match is sought between the colour of a small chip of paint and the corresponding colour on the colour wheel. Many thousands of colours have been given names and could be positioned somewhere in the wheel, so unless there is method in the search, the matching process might take some time.

What options exist for conducting this search for a colour match? Figure 2 illustrates several possible approaches, in which the crosses show the points visited during a search and the circle shows the actual colour of the chip.

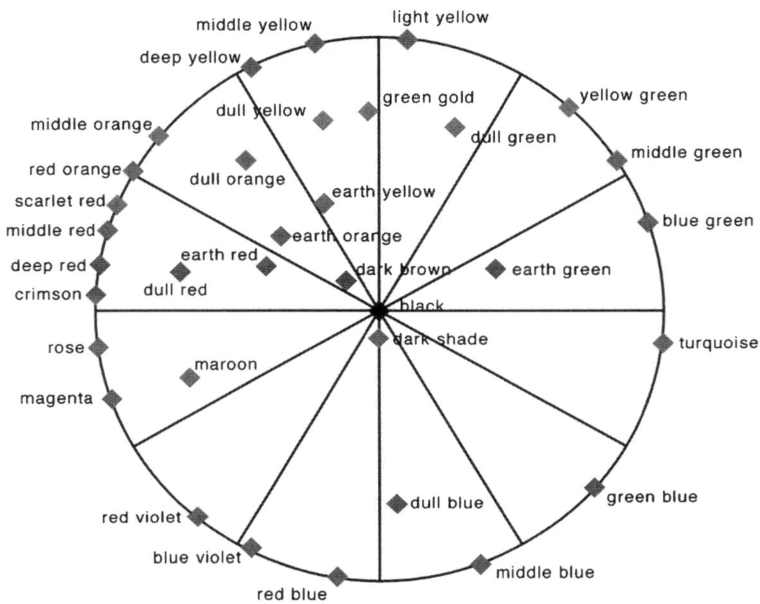

Fig. 1 A colour wheel

The dumbest of these search methods by some distance is one in which colours are chosen for matching completely at random. In a random search (Fig. 2a) a colour is chosen blindly from the wheel for comparison with the paint chip. If the colours match, the search is over, but if there is no match, another colour is chosen at random and the process is repeated. This kind of search has almost no merit beyond simplicity. Even in the small search space of this problem it is inefficient, while for large spaces the probability of success falls to near zero. Furthermore, the search may revisit points it has already checked, which is unproductive. (This is avoided in a TABU search, in which a tally is kept of points previously visited, but maintaining a TABU list adds computational overhead which becomes increasingly onerous as the search progresses and cannot pretend to rescue what is a very poor method.)

In a random search there can be no assurance that the correct match will ever be found, unless an infinite amount of time is available for looking. By contrast, an exhaustive search (Fig. 2b), in which each point is checked systematically, can at least provide this guarantee, but is still inefficient, except when the search space is small, or on those occasions when the required colour lies close to the starting point of the search.

A human searcher would be unlikely to adopt either a random or an exhaustive strategy, but would instead take into account what has been discovered about the points already inspected in an effort to speed up the search. In an elementary form this can be emulated by a colour-smart hill-climber, which moves one step at a time in the direction which most effectively reduces the difference in colour between the present location of the search and the colour of the paint chip. De-

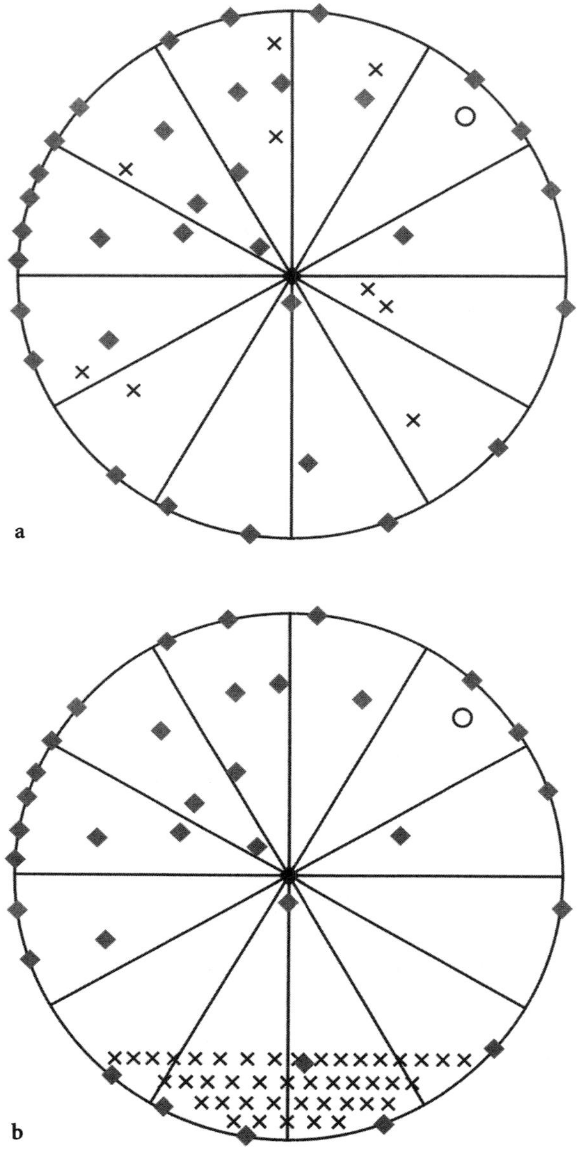

Fig. 2 Searches across a colour wheel: **a** random search; **b** exhaustive search; **c** hill-climbing; **d** evolutionary search

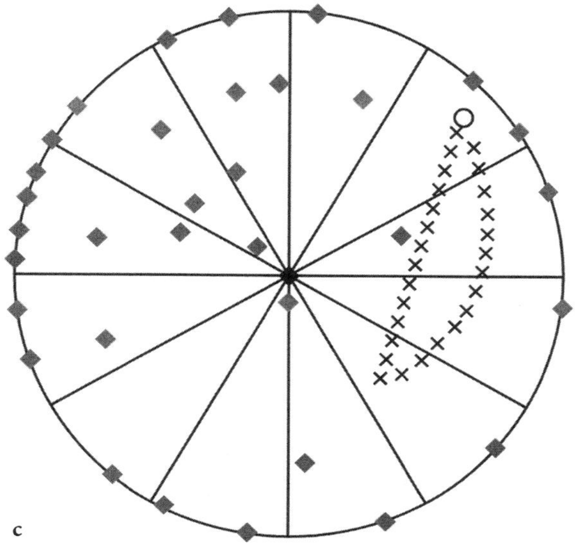

c

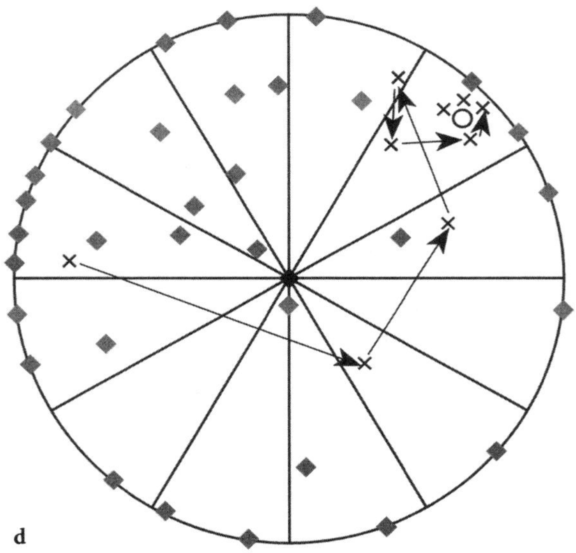

d

Fig. 2c, d (continued)

pending upon how this difference is calculated, the result may be a curving or a straight path towards the correct colour (Fig. 2c). This search is a little smarter than the previous two and, consequently, all of its moves take the search in approximately the right direction. (Although hill-climbers solve maximization problems, they can, of course, equally well be used for problems in which the minimum of a function is sought, by inverting the function which defines the search landscape, or making some other suitable transformation.)

For many types of surface the simple hill-climber is a better option than either a random search or an exhaustive search, but its limitations are not hard to spot: if the point at which it begins (which is randomly selected) is far from the correct colour, many steps are needed to bring the search into the right region of the wheel. More seriously, not every surface is as benign as a colour wheel, in which there is a smooth gradation of properties across the wheel. If the wheel were not everywhere smoothly graded, but were broken into sections, each of which contained a group of green colours, some blue colours and so on, the search could get trapped if it started in the wrong region. The hill-climber might reach a local maximum where it found a close, but not perfect, match in colour to the paint chip, but be unable to proceed any further towards the exact match if that lay some distance away. Most large-scale surfaces in chemical problems contain multiple maxima at which a hill-climber can readily be trapped, so this difficulty must be addressed by any effective search algorithm.

The question of how to deal with surfaces which exhibit multiple maxima was one factor which encouraged the development of simulated annealing. However, in contrast to simulated annealing, evolutionary approaches allow one not only to deal with the problem of multiple maxima but also to increase substantially the efficiency of the search by investigating many regions of the search space simultaneously.

Any intelligent search, such as an EA, will also take note of the closeness of the match between the paint chip and the chosen colour. If the match is good, the correct solution may be close at hand, so a small move would be wise. On the other hand, if the match between paint chip and sample is poor, the search is currently in an unpromising region of the wheel and a large jump to a distant part should be productive (Fig. 2d). Even if the direction of the jump is chosen at random and the initial jump takes the search to a second unpromising region, it may not require many such large jumps before the search arrives in a more promising area.

This type of search, in which information provided by points previously visited is used to provide clues to where to look next, is far more viable than random or exhaustive searches for moderate to large search spaces. The fact that evolutionary search combines a random element which allows it to move directly to distant parts of the search space with mechanisms to investigate local regions in detail, and that it runs several searches in parallel to provide broad coverage of the search space, gives it great power.

The combination of previous experience with a willingness to make bold jumps across the search space resembles the way that a human would work. The algorithm is, in a simple fashion, learning about the search surface, and using this knowledge in planning the next move. Whether or not this procedure will be successful depends upon a simple but critical assumption: that solutions of similar quality are clustered in the search space. If the very best solutions are scattered at random through the search space, any algorithm will struggle to find them. Fortunately, however, the type of clustering which is apparent on the colour wheel is not uncommon, being evident in a range of scientific problems.

Clustering of high-quality solutions in particular regions of the search space is an important factor in the success of the algorithms discussed in this volume – their search has a far greater chance of success if it can move towards those regions of the search space where the best solutions are located, on the presumption that, if a few good solutions are to be found there, further solutions of even higher quality may lie close at hand. Once in the right region, the search can become more focused. Just as EAs concentrate their efforts in regions with an abundance of good solutions, so they rarely return to regions of search space which appear to contain only solutions of poor quality, and in this way the computational efficiency of the algorithm is maximised.

The techniques discussed in this volume display some unexpected characteristics in addition to their ability to learn. Most notably, they have a constant thirst for random numbers. One might wonder how a program which is driven by random numbers could possibly discover meaningful solutions to problems. We shall see shortly how random numbers play their crucial role.

3 Evolutionary Algorithms

3.1 Change and Competition

EAs differ from conventional search algorithms in several ways. As the name suggests, EAs incorporate mechanisms which are related to the operations of natural evolution. The driving forces for evolution are the creation of progeny which contain parental genetic material, the susceptibility of this material to modification through mutation, and the effect on progeny of selection pressure. We can anticipate that analogous forces will operate within EAs. Although the algorithms display evolutionary characteristics, they are not inherently evolutionary; their evolutionary character is not hardwired into them, but instead becomes apparent as they run, in a fashion known as emergent behaviour.

Evolution is ineffective when it works upon a single individual. Although random mutations will cause a gradual drift in the genetic material of a succession of individuals, there is no selection pressure which can promote the survival of better individuals at the expense of the less well-adapted (short of the extreme pressure that causes the death of the individual without leaving issue, when, inevitably, evolution comes to an end). Any evolution that occurs within a population of 1 will as a consequence be merely a random walk which makes progress only by chance and is as likely to delete a productive mutation as retain it. If evolution is to lead to a gradual accommodation of individuals to the challenges posed by their environment, some selection pressure is required so that poorer individuals, which are less able to cope with the environment, are slowly squeezed out by their stronger competitors.

Competition unavoidably requires a population size greater than 1 – a single individual cannot compete with itself. Since EAs show evolutionary behaviour, it is reasonable to anticipate that they too will normally need to work upon a group of individuals. This requires that these algorithms must operate on many potential solutions simultaneously, so that selection pressure can be applied to cull the poorer solutions and drive the search towards those of higher quality. This manipulation of a group, or population of solutions, is a fundamental difference with most other optimisation methods, which typically create and then refine a single solution.

Individuals are the units upon which natural evolution operates, and also the unit manipulated in an EA, in which each individual is correlated with a distinct solution to the problem being studied. These individuals may be a direct representation of the solutions themselves in numeric or symbolic form, a list of atomic coordinates, for example, or they may instead be a coded form of that solution. Individuals are processed using evolution-like operations, the role of which is to gradually transform them from initial randomly chosen, and probably poor, solutions into optimum solutions.

3.2 A Simple Evolutionary Algorithm

The steps which constitute a simple EA are outlined in Fig. 3.

We can illustrate the operation of an idealised algorithm using a design task. Suppose a model EA sets to work constructing a lighthouse from a kit of cylindrical building blocks, a small fraction of which contain light sources. We already know what a working lighthouse looks like, so it should be easy to judge whether the algorithm makes progress.

3.2.1 *Initial Population*

A population of individuals is required for the EA to work on, so let us begin by making that. A portion of our random starting population of lighthouses is shown in Fig. 4, in which light-emitting blocks are shown cross-hatched and structural building blocks are shaded.

None of the lighthouses in Fig. 4 will win awards, but we must still in some way assess their quality so that the algorithm can distinguish between them. The yardstick of quality for each individual is known as its fitness. This may be measured by the comparison of one individual with others in the same population, or by the extent to which the solution represented by an individual is able to satisfy some external requirement, such as being able to reproduce an experimental spectrum or, in this instance, to function effectively as a lighthouse.

The way in which fitness is determined is problem-dependent. A lighthouse must, first and foremost, produce light; it must do so at an appropriate distance

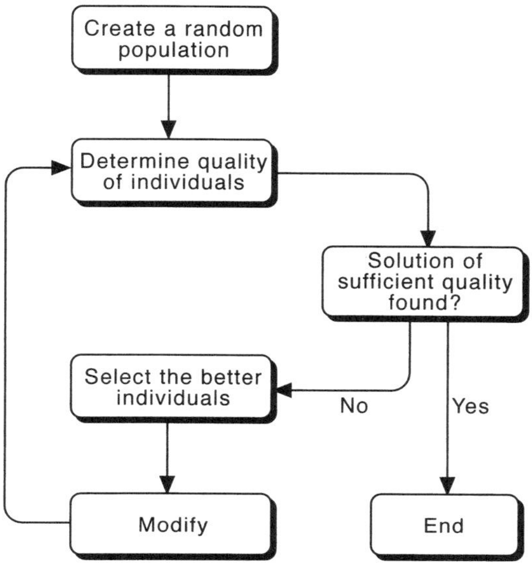

Fig. 3 A simple evolutionary algorithm

above ground, it must be stable, not excessively costly to build or too large for the site, and so on. Taking into account factors such as these, we can calculate a fitness for each structure; these values are shown in Fig. 4. The procedure used to determine fitness is entirely a matter for the user – it is not prescribed by the EA. It is therefore simple to bias the calculation to encourage the development of solutions which exhibit a particular characteristic; we could place a premium on the ability of the lighthouse to emit light, or on the stability of the structure if we wished, by weighting those factors more heavily than others in the calculation of fitness.

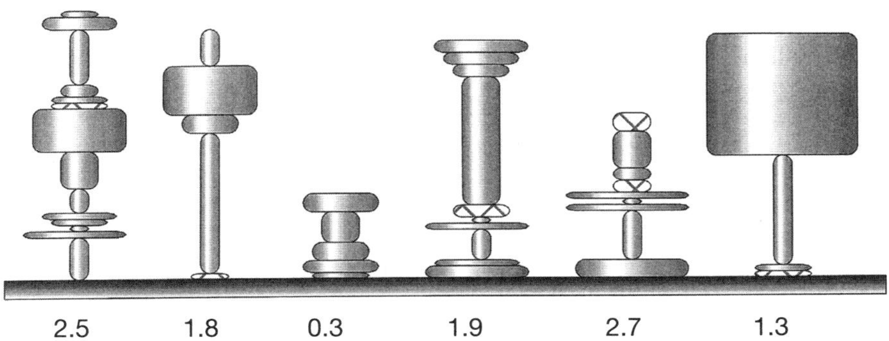

Fig. 4 A portion of the lighthouse population, generation 1. The *figures* indicate the fitness of each design

3.2.2
Selection

It is now time to start the evolutionary clock running. To create a new generation of individuals, selections are made from among the members of the present population, copying chosen members one by one into a new population. Survival of the fittest dictates that, in selecting individuals to form part of the next generation, we should choose preferentially the better solutions. Members from the first population which are picked to be placed into the next one are parents; their progeny are referred to as children.

A variety of methods, stochastic, deterministic or a hybrid of the two, exist for selecting parents; we shall outline some of them later. For the present, let us assume that some suitable stochastic selection method has been applied, in which the better solutions are more likely, but not certain, to be picked, and a new population is formed. This will be of equal size to the starting one, and since survival of the fittest promotes proliferation of the more promising designs at the expense of the poorer, we anticipate that the average fitness of the new population will be as good as, or superior to, that of the parent population. (Recall, though, that selection may contain a stochastic element; because of this it is not certain that the average fitness will improve from one generation to the next.)

But although the average fitness may improve, there can be no designs yet in the new population that were not present before, since we have merely copied designs from one into the other. To make progress, a way must be found to modify some individuals so as to create new designs. There are various ways in which this may be done; we shall consider the two most fundamental: a cut-and-paste operator and mutation.

3.2.3
Swapping of Genetic Material

Cutting-and-pasting does what is says on the label. Portions of two individuals are cut out and swapped to create two new children. The individuals involved and the cut points of the segments to be swapped are all chosen at random. Following cutting-and-pasting, the two children will each inherit some of the characteristics of both parents.

There is, of course, no reason to suppose that either of the new individuals resulting from this operation will necessarily be fitter than their parents. However, if good segments from two different parents are, by chance, brought together in a single child, it is probable that the child will have superior fitness; it will then be likely to survive for at least several generations, so the positive characteristics provided by the parents will be retained in the population and may multiply as the generations pass. Conversely, if the child is of low fitness because it contains few features of value, it is travelling up an evolutionary dead end, and will before long be culled by the selection process.

3.2.4 *Mutation*

When cutting-and-pasting is complete, every individual in the new population is offered up as a candidate for mutation. In this problem our mutation operator will select a building block at random and replace it by a fresh block of randomly chosen design.

Although mutation is usually applied infrequently and many individuals in the new population will escape unscathed, this is still a crucial operation. It is important that diversity be retained in the population, but the selection operator militates against this. Poorer solutions are weeded out by selection, leaving the better solutions to proliferate, and this can lead to the population gradually homogenising. Ultimately, the population fills up with many almost identical individuals, and any advantage which might be provided by working with a population will be lost. Occasional random mutations help to reduce the likelihood of this stagnation.

Mutation is also important as a source of new information. In the present example, if mutation were not included, the form of the lighthouses would be limited to those designs which could be constructed using only the particular set of cylindrical blocks which were present in the first population. This serves to place a cap on the number of different lighthouses which are possible. In the worst-case scenario, if none of the initial sets of designs included a source of light, all lighthouses would forever be dark unless a mutation operation existed which could create a light-emitting segment.

Once selection, crossover and mutation have been applied, the new population is complete.

The mutation and crossover steps muddy the picture a little, but it appears from the second generation (Fig. 5) that neither of the two poorest designs from generation 1 have made it through the selection process. In particular, the small dark lighthouse from population 1, which was clearly an also-ran in the evolutionary stakes, has been eliminated, and all the designs, while still of little architectural merit, are at least functional. The evolutionary steps are now repeated,

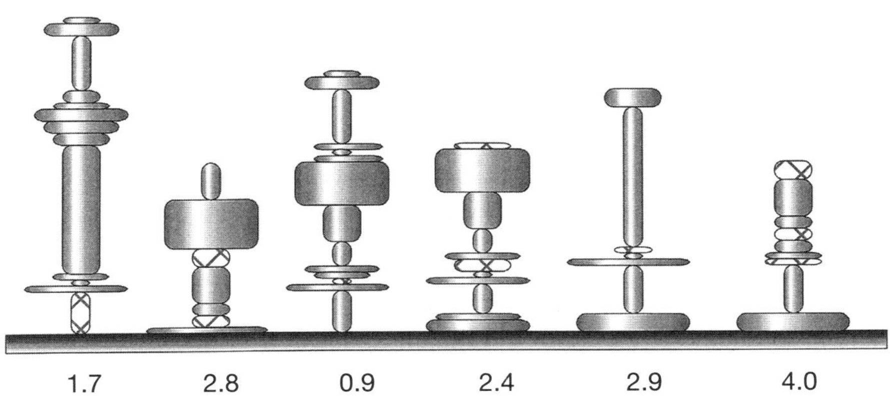

Fig. 5 A portion of the lighthouse population, generation 2

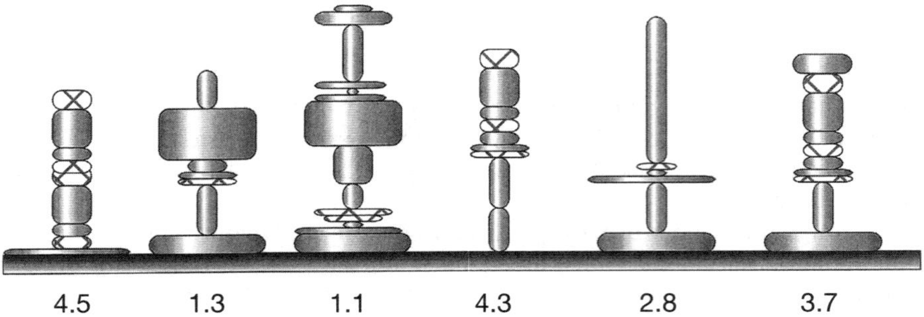

Fig. 6 A portion of the lighthouse population, generation 3

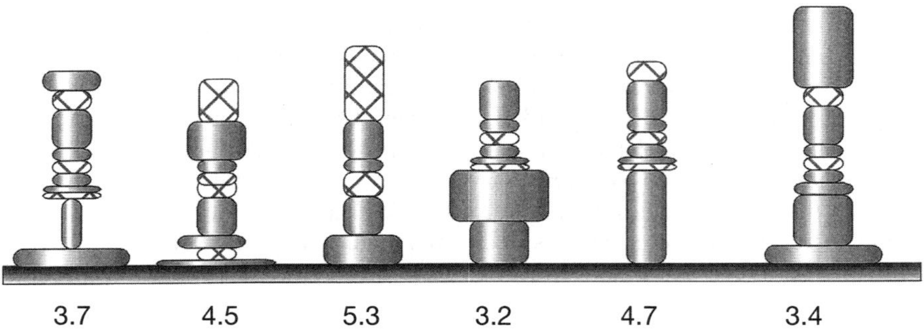

Fig. 7 A portion of the lighthouse population, generation 10

the fitness of each new design is determined and used to select a further set of individuals. The new child population is treated with the cut-and-paste and mutation operators and the process continues.

A representative portion of the third generation is shown in Fig. 6, where a gradual improvement in the population is becoming evident. By the tenth generation (Fig. 7) the designs are becoming more realistic, and the features to be expected in a working lighthouse are starting to emerge.

3.3 Population Drift

The sharing of good design features among individuals and the existence of selection pressure to filter out the poorer designs encourages a progressive drift of the population towards regions of the search space in which the better designs are clustered, that is, areas rich in those designs which meet the criteria used to determine fitness. Ultimately, by focusing most of the search effort on these regions, the algorithm may discover optimal solutions. It is, in fact, the individuals themselves that move across the search space. Just as the best designs are clustered in certain regions of that space, so, as the algorithm runs, individuals gravitate towards promising areas.

EAs repeatedly carve up old members of the population to create fresh solutions. As in natural selection, competition within the population is essential, otherwise its evolution would be unpredictable and undirected, the algorithm would be as likely to retain poor solutions as promising ones and would make a lengthy and probably unproductive random walk over the search surface. Since individuals in the current population have evolved from those created in past generations, they reflect some of the lessons learned during previous attempts at solution. It is in this fashion that the algorithm learns about a problem.

The parallel with nature, evident in the application of a survival of the fittest strategy, the crossover and mutation operators, and the gradual improvement in the characteristics of the population as a whole is appealing, but is only partial. In the natural environment, a population grows in size exponentially if conditions allow, in other words if food is unlimited and predators have a negligible impact. In an EA, the population consists of a set of candidate solutions, and the mechanism described previously does not contain an operator whose effect would be to increase the population size. It is true that, by diminishing the selection pressure and allowing a higher proportion of individuals to survive, one could arrange for the population to grow as the generations passed, but it is not clear that there would be anything to be gained in allowing the less-well-adapted individuals to survive. If the population grows indefinitely, each individual will be processed less and less frequently, and the rate at which solutions improve will fall away. In some types of algorithm, for example messy genetic algorithms (mGAs), the size of the population may be altered during the calculation, but even then the population growth is limited, and in most EAs the population size, once chosen, remains fixed.

In a further difference with natural processes, no operation analogous to mitosis (cell division) exists in EAs, since the key aspect of the operation of the algorithm is the way in which the gene pool within the population changes from generation to generation, not the fate of a particular individual.

3.4 Evolutionary Variables

We noted earlier that conventional iterative methods often contain adjustable parameters, the values of which had to be chosen with care, since they would influence the efficiency and success of the calculation. We have not escaped from these parameters by using an EA, and a set of variables is emerging whose values will affect the course of the calculation. There is the population size – how should that be chosen? At what rate should individuals be mutated? Should child solutions be selected deterministically or stochastically – or using some hybrid method?

In view of the importance of choosing appropriate values for EA variables, it is common for these to be selected through a combination of past experience with similar problems and some testing of the dependence of the quality of solution reached with different population sizes, mutation rates and so on. Since the population size in different methods varies considerably (in a GA it is typically in the range 30–100, while genetic programming (GP) populations may be hundreds or thousands of times larger) a reasonable value for this parameter is vital.

The EA methods discussed in this volume are computational tools. They are neither magic nor impenetrable; instead, they are recipes which lay out a strategy for solving a problem, so in this sense they are analogous to conventional methods. They provide a route which is increasingly being followed by those tackling problems in science which have been found to be hard to solve using other techniques. We outline in the following the principles of three of these methods.

4 Genetic Algorithms

The GA is the most widely used EA within chemistry. There is a substantial, and growing, literature which demonstrates the power of this method as an optimisation tool within not just the chemical sciences, but across a range of scientific disciplines. Provided that the problem is of the appropriate structure, it is a very powerful optimisation method. It is also tolerant, in other words it is capable of providing useful solutions even when values of variables governing its operation, such as the population size, are not chosen optimally.

4.1 Strings – the Genetic Algorithm Solution

There are limits to the applicability of virtually every optimisation method. A GA may only be applied to a problem if it is possible to express the solution as a sequence of values. This sequence is referred to as a chromosome or a string, and each parameter within the chromosome is a gene. The entire set of genes constitutes the genotype, and the solution to which this genotype corresponds is known as the phenotype (Fig. 8). The GA works to refine strings with the help of the evolutionary operators outlined previously.

Examples of typical GA strings are as follows:

$$S_1 = \{0110101000101011011100....\}; \tag{1}$$

$$S_2 = \{2.377, 1.76, -2.005, 17.118, ...\}; \tag{2}$$

$$S_3 = \{ABGTFDAGRFRRIHKD....\}; \tag{3}$$

$$S_4 = \left\{ \begin{bmatrix} 4 & 11 \\ 8 & -1 \end{bmatrix}, \begin{bmatrix} 2 & 3 \\ -4 & 0 \end{bmatrix}, \begin{bmatrix} 0 & 0 \\ 1 & -9 \end{bmatrix}, \dots \right\}. \tag{4}$$

The strings represent the solution to the problem, generally in coded form, and the algorithm makes a search through this coded search space (the genotype space). The string might represent a calculated IR spectrum, the order in which different chemicals are fed into an industrial reactor, a sequence of temperatures, or the molecular constants which define the geometry of a molecule.

GA strings are generally one-dimensional, but two-dimensional and higher-order strings have also been used where such a representation permits more

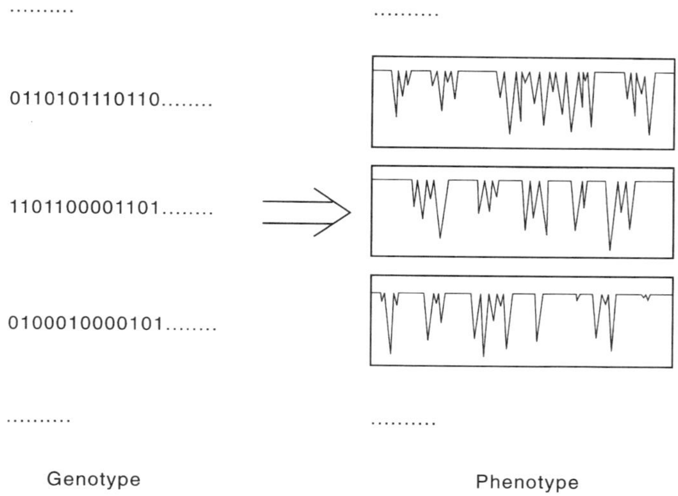

Fig. 8 The relationship between the genotype (coded solution) and the phenotype (actual solution).

efficient calculation. The genetic operators which are discussed later must be modified for use with strings of higher dimensionality, but although computationally somewhat more demanding, multidimensional strings present no significant additional problems in manipulation.

The requirement that it be possible to represent the solution as a string is a necessary but not sufficient requirement for successful use of the GA, but when the problem has the appropriate structure, GAs may outperform conventional methods of solution by a significant margin. The characteristics that problems must display if the GA is to be an appropriate method to tackle them are discussed in most standard texts (see Further Reading).

In early GA work, the strings were almost exclusively binary (S_1), and within computer science, binary strings are still common For some types of problem it may be appropriate to use real-valued strings (S_2), in which each entry represents part of the actual solution in uncoded form. A binary string can, of course, be interpreted as representing a real number, but if it is simplest to express the solution to a problem as a sequence of floating-point values, binary representation introduces a computational overhead which may slow the calculation. As scientific problems are more commonly expressed in terms of floating-point numbers than binary numbers, real-valued strings are now common in chemistry. There is a secondary advantage to the use of real-valued strings; a sequence of real numbers is more readily interpreted by eye than a binary string (unless the solution is naturally cast in binary format), and this can be helpful in monitoring an ongoing calculation.

Other choices, beyond that of choosing binary or real-valued genes, may be available in the way in which the solution is represented and it is important that this be considered at an early stage. Just as the geometry of a molecule might be specified as the coordinates of each atom (in Cartesian or polar form), as a connectivity matrix, or as a set of dihedral angles, bond angles and bond lengths, so each of these types of parameters could be used to construct a string which represents the molecule. The success of the GA is strongly related to the form of representation chosen, and it is not unusual for one representation to lead to impressive results, while another gives poorer results, or results of no value, so the choice of an appropriate way to code solutions is a crucial step.

4.2
Genetic Algorithm Operators

The operations which constitute a simple GA are shown in Fig. 9.

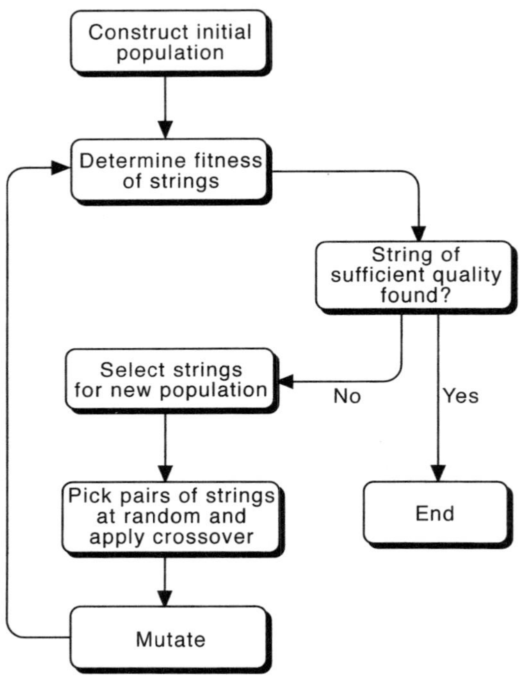

Fig. 9 The genetic algorithm

4.2.1
Initial Population

Initially a population of random strings is created; domain-specific knowledge if available may be used to ensure that the values (known as alleles) which are chosen for the genes lie within appropriate limits. For example, if the string is to represent an IR transmission spectrum, all initial values should be chosen from within the range 0–100%, since a transmission outside these limits is physically unreasonable.

4.2.2
Fitness

The fitness of every string is then determined, after the genotype has been decoded, if necessary, into the corresponding phenotype. The fitness measures the extent to which the solution satisfies the requirements of the problem. If the string represented the geometry of a trial molecule and an experimental microwave spectrum of a real molecule, whose structure was sought, was available, we might gauge the fitness of the trial molecule encoded by the string by the degree to which the calculated microwave spectrum of the molecule agreed with the experimental spectrum.

4.2.3
Selection

Once the fitness of each string has been found, selection follows. This may be deterministic (the better of two possible solutions is always retained, while the other is invariably discarded), stochastic (the best of two or more solutions is chosen with a probability related to the fitness of the solutions), or a combination of these.

A variety of methods have been proposed for the selection step. One of the earliest, and most straightforward, is the roulette wheel, in which the probability of a string being chosen, p_i, is proportional to the fraction of the total fitness of the population which that string represents:

$$p_i = f_i / \sum_j f_j . \tag{5}$$

On the roulette wheel, each string is allocated a segment whose size is proportional to the string's fitness (Fig. 10). The wheel is spun, and the string into whose slot the imaginary roulette ball falls is selected to be copied once into the new population. The process is repeated until the new population is equal in size to the starting population.

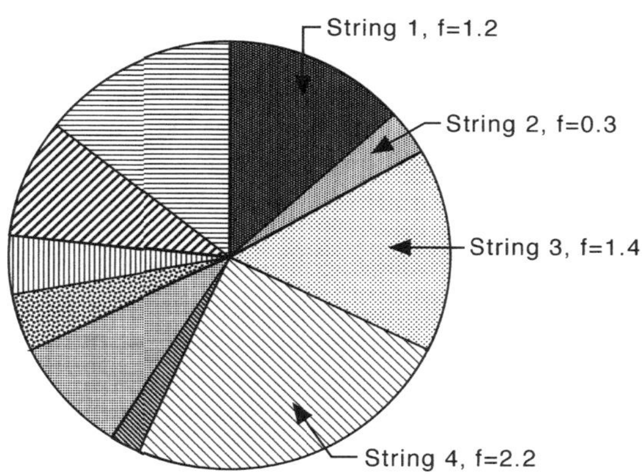

Fig. 10 Roulette wheel selection

Table 2 Application of stochastic remainder selection

String	Fitness	Normalised fitness	Strings sure to be made	Residual fitness	Strings from roulette wheel
1	1.73	0.88	0	0.88	1
2	3.11	1.58	1	0.58	1
3	0.88	0.45	0	0.45	0
4	0.78	0.40	0	0.4	0
5	2.54	1.29	1	0.29	0
6	0.75	0.38	0	0.38	1
7	3.95	2.01	2	0.01	0
8	2.39	1.21	1	0.21	0
9	1.66	0.84	0	0.84	1
10	1.89	0.96	0	0.96	1

This procedure has the advantage of simplicity, but since it is stochastic there is no certainty that the best string in the parent population will be selected to be copied into the new population. If, once a good string is found, it is lost from the population during selection, convergence may be slowed. The risk of loss of the best string becomes more marked as the population homogenises and the fitness values of the strings become more alike (although in compensation the damaging effect that loss of the best string has upon the search diminishes since many close relatives of it may remain in the population, so much of the genetic material will be retained). A modified procedure, known as stochastic remainder, addresses this difficulty (Table 2).

In stochastic remainder, the fitness values of all N strings are first normalised, so that the average fitness is 1:

$$F_i = f_i \times N / \sum_j f_j . \tag{6}$$

Each string is then copied into the new population a number of times equal to the integer part of the fitness (so two copies would be made of a string whose normalised fitness is 2.01, none of a string of normalised fitness 0.88 and so on). The number of copies made in this way is subtracted from the normalised fitness and a roulette wheel is run to fill up remaining slots in the new population using the residual fitness values. This procedure guarantees at least one place in the new population for every above-average string, but retains the element of chance that gives even strings of quite modest quality the possibility of a place in the next generation.

Further procedures, such as tournament selection (in which pairs of parent strings are chosen at random and the string of higher fitness is copied into the child population), are also widely used and are discussed in standard texts on the subject.

4.2.4
Crossover

Once sufficient individuals to fill the new population have been chosen, strings are drawn in pairs from this population for exchange of genetic information. Just as there are several different ways in which selection can be carried out, so the information exchange can take place in a variety of ways. In a simple two-point crossover step, pairs of strings are chosen at random and cuts are made at randomly selected positions. The segments between these cuts are swapped. This has the effect of mixing the genetic material between the two strings (Fig. 11).

All genes in the string should have an equal chance of being processed by crossover. If a two-point cut is used to snip out a random portion from the middle of a linear string for exchange, the ends of the strings are sampled less frequently than the centre. While fresh material is readily incorporated into the centre, the outer segments may be modified only infrequently. This can be remedied using a wrap-around operator which samples all genes with equal probability (Fig. 12).

The transfer of material from string to string which is brought about by the crossover operator parallels sexual reproduction. Through this mechanism portions of good solutions which have developed in different strings can be brought together. However, this conjunction of good portions is still a matter of chance, since the process involves strings chosen completely at random. No attempt is made to identify good strings and deliberately bring them together for crossover.

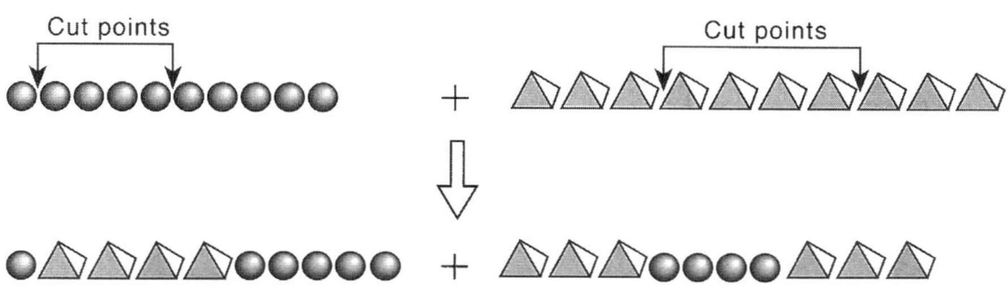

Fig. 11 Simple two-point crossover

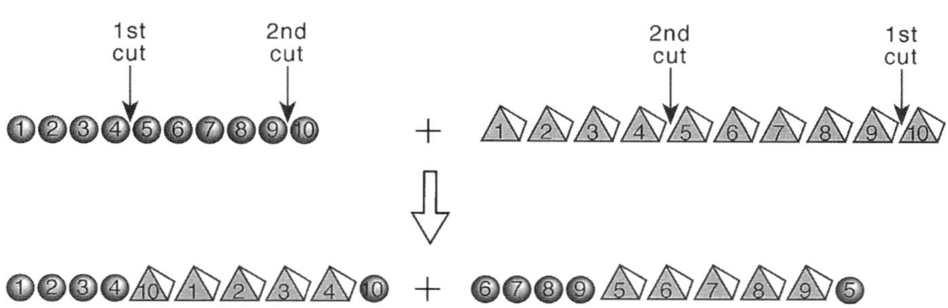

Fig. 12 Wrap-around crossover

Some types of strings may need to be repaired after crossover. Suppose each string represented the order in which a set of C chemicals should be fed into an industrial reactor; every string would then contain each of the values $1, 2, \ldots C$ exactly once. After crossover it is probable that some chemicals would appear twice in one string and be missing from the other. This could be dealt with either by a repair step following crossover, or by using a crossover operator constructed in such a way that only crossover operations which lead to valid strings are permitted.

4.2.5
Mutation

When crossover is complete, a small fraction of strings are mutated to introduce new genetic material. A mutation may consist of the replacement of a randomly selected gene by a fresh random value, the swapping of the values of two randomly chosen genes within a single string, a gentle nudge to the value of a gene in a randomly selected direction, or some other process. Mutation rates are usually low (typically of the order of 1 per 1,000 genes) because while mutations introduce new, possibly crucial information, they also destroy potentially valuable information which has evolved during the search.

A fine balance must be maintained between the need to create and enhance the diversity and to minimise the damage caused by loss of information. In natural evolution, mutation rates which are very low lead to only gradual change in the characteristics of a population, and it may then be unable to adapt with sufficient speed to changes in the environment. At the other extreme, high mutation rates cause destruction of information and the risk that new individuals may not be viable because of the loss of vital genetic data. The optimum conditions for evolution in both natural and artificial systems seem to be those in which mutations occur at a rate which is just below that at which significant amounts of information are lost.

The roulette wheel and tournament selection methods may lead to accidental loss of the best string in the population which is undesirable. In addition, even if the best string is selected for insertion into the new population, it may still be altered by the genetic operators. In order to protect the best string from damage by crossover or mutation it is common to employ an elitism strategy in which one copy of the best string is protected from attack by the evolutionary operators, or is given a lifetime that exceeds one generation.

The new population is now complete and the cycle continues, with the fitness of each member of that population being assessed in the same way as before.

A variety of further steps, beyond simple measures such as elitism, can be taken to enhance the performance of GAs. These include breaking the population into several subpopulations, using local knowledge, periodically doing local searches using hill-climbing or similar techniques starting at points identified by the GA as promising, and using string fridges (occasionally inserting strings extracted from previous generations to enhance the diversity of the population).

5 Evolution Strategies

Evolution strategies (ESs) were developed in the 1960s, at around the same time as early work on GAs. In its first form, ES was not really an EA, more a slightly disguised, moderately sophisticated hill-climber; however, it has since metamorphosed into a true EA through the use of a population of potential solutions and a more sophisticated evolutionary mechanism.

In an ES the phenotype and the genotype are the same. Solutions to the problem are thus represented directly as numbers which the algorithm manipulates, in contrast to the GA, in which the algorithm manipulates a genotype representing the solution in a form which may be encoded. In this respect ESs are somewhat more transparent than GAs.

In its early incarnation the algorithm was slightly evolutionary, but very unsociable. It resembled a GA which used an elitism strategy, working with a population of 1. Although this preliminary form is now rarely used, it is helpful to outline its mechanism since it illustrates most elements of the modern, more effective algorithm.

5.1 Basic Evolution Strategies

The basic ES operates upon just one individual which is modified in an iterative process through the following operations:

1. An initial candidate solution (the first parent) is created either entirely at random or, as in the case of the GA, by using available domain-specific information about the solution space.
2. A child solution is created by mutating all values in the parent in the manner outlined later.
3. The fitness values of parent and child are found by testing their quality in the normal way. If the child is fitter than the parent, or is no worse, the child replaces the parent, and the parent is discarded; otherwise the child is discarded and the parent survives.
4. This process is repeated from step 2, starting with the surviving individual, until a solution of suitable quality is discovered or the maximum number of iterations is reached.

This process is a two-numbered or two-membered ES, since just two individuals – the parent and the child – compete to survive each generation. Any standard ES may be specified in shorthand of the form (μ, λ)-ES or $(\mu+\lambda)$-ES, where μ and λ are the numbers of parents and children, respectively, so the scheme just outlined is a (1+1)-ES. The difference between the two types of notation (μ, λ)-ES and $(\mu+\lambda)$-ES will become apparent shortly.

The parallel with a minimalist GA, in which the better string of parent and child within each generation survives to the next generation, is obvious, but while the population size is a clear difference between GAs and this elementary form of the ES, there are also more fundamental differences. Unlike the GA, the ES chromosome contains at least two types of information. There is, of course, the set of values of the parameters which define a potential solution, but the chromosome also contains strategy parameters which determine how the algorithm will operate. Each string is thus made up of two parts:

$$V = \{\mathbf{x}, \boldsymbol{\sigma}\}, \tag{7}$$

where $\mathbf{x}$ is a point in search space, constructed as a real-valued vector and $\boldsymbol{\sigma}$ is a real-valued vector of strategy parameters which is used to direct the evolution of the algorithm. In the simplest ES, $\boldsymbol{\sigma}$ consists of a group of standard deviations equal in number to the length of the vector $\mathbf{x}$. A typical ES individual might consist of a set of genes such as

$$\{3.4, 2.51, -8.22, 0.00, 9.11, 27.8, 0.02, 0.001, 0.11, 0.15, 0.3, 0.06\},$$

in which the first six numbers comprise the putative solution, and the second set of six values is an ordered group of standard deviations; the standard deviation 0.02 is associated with the parameter value 3.4, and so on.

The standard deviations within $\boldsymbol{\sigma}$ determine the size of the changes which are permitted when the mutation operator, which is the principle method by which solutions evolve, is applied. Mutation is the core of ES, and, by contrast with the GA, causes modification of the entire phenotype every generation. During mutation, each value in the vector $\mathbf{x}$ is modified by the addition of a positive or negative random number, whose size is related to the appropriate standard deviation.

$$x_{i,t+1} = x_{i,t} + N\{0, \sigma_i\}, \tag{8}$$

where $x_{i,t}$ is the value of the ith parameter in $\mathbf{x}$ at generation t, and $N\{0,\sigma_i\}$ is a random number chosen from a Gaussian distribution with a mean of zero and a standard deviation of σ_i. A Gaussian distribution is used to reflect the fact that in natural evolution small changes in a chromosome are more likely than large changes.

In early ESs the vector of standard deviations $\boldsymbol{\sigma}$ was kept constant throughout the calculation. It can be shown that, if $\boldsymbol{\sigma}$ is constant, the algorithm will converge to an optimal solution, but the time required for convergence is not defined. To encourage rapider convergence, the strategy parameters which direct the mutation operations may themselves be mutated using an update rule during a run.

Once mutation has been performed the fitter of parent and child is selected and the process repeats. There can, of course, be no possibility of applying a crossover operator when the complete population consists of just one solution, so mutation is the only genetic operator available.

5.2 Population-Based ES

An ES operating upon a single individual is prone to getting stuck on a local maximum, leading to a premature end to the calculation. It was recognised that an ES would have more chance of avoiding being trapped at local maxima if it were to operate on a population whose size was greater than 1.

At the start of a population-based EA, potential solutions are once again created at random. Alternatively, they may be produced by starting from user data which define roughly the search space within which a solution is sought, and then making random mutations to this to create the initial population.

As the population now contains several parents, crossover becomes possible, and randomly chosen individuals can be selected for treatment by recombination operators. All members of the population have an equal chance of being selected for this crossover, in contrast to a GA, where the string fitness determines the probability that a string will get through to the mating stage. ES crossover operators may function like crossover operators in a GA, bringing together information from two parents, or may be more broadly based, combining data from several parents. Once crossover is complete, the new solutions are treated with the mutation operator, selection is carried out, and the cycle repeats.

As a population of solutions now exists, two variations are opened up which we can use to select members of the next population. In both methods, the members of the child population are chosen deterministically, in other words, through a type of tournament selection in which the better of two solutions always wins.

1. ($\mu+\lambda$): Select μ parents and λ offspring. Mutate all individuals, then select the best μ as parents for the next generation.
2. (μ, λ): Create λ offspring from μ parents and allow only the offspring to compete for survival. It is essential that λ is greater than μ in order to provide some form of selection pressure. All parents are replaced by offspring in this scheme.

Much effort has been devoted to investigating methods by which convergence in the ES can be encouraged. Rechenberg proposed the "one fifth success rule" which, though simple conceptually and in implementation, is still effective in most cases. It suggests that if the ratio of successful mutations to all mutations is less than 0.2, the variance of the mutation operator should be decreased, while if the ratio is greater than 0.2 the variance should be increased:

$$\begin{array}{lll} \text{if} & \phi(k) < 0.2 & \text{then } \sigma = \sigma \times c_d; \\ \text{if} & \phi(k) > 0.2 & \text{then } \sigma = \sigma \times c_i; \\ \text{if} & \phi(k) = 0.2 & \text{then } \sigma = \sigma. \end{array} \tag{9}$$

k is a parameter to the algorithm and specifies how many cycles should elapse before the one-fifth rule becomes operative, $\varphi(k)$ is the fraction of mutations in generation k which lead to an improved solution, and $c_i > 1$, $c_d < 1$.

The purpose of this rule is to try to ensure that the algorithm always moves towards the optimum solution at a suitable speed. If it is found, at some point in the calculation, that mutation is frequently leading to improved solutions, larger jumps should be used to try to move more rapidly towards the optimum solution. On the other hand, if most current mutations are unproductive, the optimum may be close, so large steps are likely to move the search further away from the optimum, and the search should be narrowed by using smaller steps.

6 Genetic Programming

6.1 Hands-Off Programming

Automatic programming is the task of constructing and running an algorithm which can create and test its own computer programs. In its idealised form, GP generates programs in an entirely automatic fashion to solve problems posed by the user.

At first sight, GP appears to be some sort of holy grail. It addresses one of the central issues in programming – how can a computer program be made to do what is required without being told exactly how this must be done? In GP the computer is not given detailed instructions on how to accomplish its task, but builds and tests computer programs from scratch without intervention from the user. It needs no knowledge of the underlying structure of the problem that is to be solved. In principle therefore, a computer-illiterate user could use GP to prepare a computer program capable of solving difficult scientific problems. The only input required is a specification of the task which the genetic program is to complete.

This sounds like the ultimate computer program – a piece of software which, when pointed in the right direction and told what to aim for, can yield a complete solution to a problem. That at least is the principle.

6.2 Genetic Programming Individuals

GP was originally proposed as a means by which intelligent machine behaviour might evolve. Genetic programs are a variation of the GA in which each string is no longer a linear sequences of values, but a list of instructions for preparing a computer program whose role is to solve the problem of interest. The GA part of the technique operates on a pool of such programs which differ in size, shape and effectiveness as problem-solvers. In a conventional GA the strings are of fixed length, but GP strings have to be flexible, since it is not possible at the start of the calculation to define how many steps (and thus how many genes) will be needed to build an effective program.

Although tht precise size of the programs is not fixed, a maximum size must be, since if no maximum is specified, programs may grow almost without limit (code bloat), as they fill up with inefficient, replicated or irrelevant code.

The genetic program starts with a population of many thousands or millions of randomly generated computer programs and uses evolutionary principles to breed new and improved programs. The strings on which a genetic program operates are interpreted as parse trees; an example of a simple parse tree is given in Fig. 13.

The set of internal nodes used in the parse trees is known as the function set F, where $F = \{f_1, f_2, \dots f_{Nf}\}$. All functions have an arity (the number of arguments) greater than 1. The set of terminal (leaf) nodes in the parse tree is, predictably,

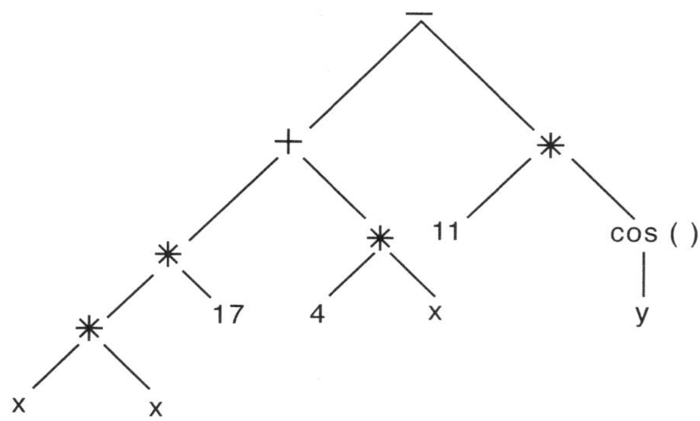

Fig. 13 A simple parse tree for the expression $4x + 17x^2 - 11 \cos y$

known as the terminal set $T = \{t_1, t_2, \ldots t_{Nt}\}$. These are the computational ingredients which the algorithm uses to construct programs.

The functions F must be able to accept as arguments any data type which is likely to be created by any other function, or created by a terminal in the terminal set. They may return one or more values when called, or a diagnostic value which specifies that an invalid operation has been attempted (for example, a request for the square root of a negative number), but whatever the arguments passed across to them they must always return a valid response. The search space investigated by the genetic program then consists of the entire set of all possible programs constructed from these functions and terminals, including recursive functions. GP has been pioneered by John Koza, who has shown its versatility in producing quite complex programs even when the permissible operators are restricted to quite simple arithmetic operations, such as addition, subtraction or multiplication.

6.3 Genetic Programming Operators

The search algorithm in the genetic program incorporates a classical GA with mutation and crossover. The evolutionary operators, under the guidance of the GA, direct the evolution of the pool of programs, using as the fitness function a measure of how effective the program is at solving the defined problem.

The steps in running a genetic program are as follows:

1. An initial population of strings is created as a kind of primordial program soup.
2. The effectiveness of each program in the soup at solving the problem is determined by running the program and comparing the actual output of the program with the desired output. Postprocessing of the output is often required and is performed using a wrapper, or output interface, whose role is to transform the output of the program into a form which can be interpreted by the outside world. Each candidate program may need to be run several times to assess its fitness so that an average fitness can be calculated, especially if there

are many sets of input data which the program should be able to handle. A termination criterion which specifies the degree to which the output of the program must match the desired output for execution to cease may be specified, or simply left undefined, with the program allowed to execute indefinitely.
3. If necessary the fitness values are adjusted through a normalisation or scaling process. A new population of candidate programs is then created by selecting programs with a probability related to their fitness. The breeding of new programs occurs in ways which will now be familiar, using a survival-of-the-fittest criterion to select those programs which are most accomplished (or, in the early stages of the calculation, least hopeless) at solving the problem.
4. Randomly selected individuals are combined by crossing over subtrees between pairs of programs, selecting the subtrees at random. Generally entire subtrees are swapped. If necessary, the code after crossover is corrected to ensure that the resulting programs are syntactically valid, though this step may be avoided if the crossover operator is designed in such a way that only syntactically reasonable changes are allowed.
5. Occasionally randomly selected programs are mutated by changing a random subtree. A subtree is chosen and replaced by an entire new randomly selected subtree, using the same procedure which created the programs at the initialisation stage. As usual, the purpose of mutation is to introduce new information and to prevent the algorithm from stagnating.
6. The algorithm resumes at step 2 with calculation of the fitness of the new programs.

The fitness measure is a key factor in the performance of a genetic program. Quite clearly, it is problem-dependent, being related to the difference between the desired and the actual output. While for most problems this difference is simple to determine, that is not always the case. For example, if the desired response is for the program to be able to direct a robot through a maze, determining a measure of success when the robot fails to get very far into the maze may be difficult. The fitness might then be measured as the length of time for which the robot behaves correctly, though it might alternatively be assessed by judging how accomplished the robot is in working through the sections of the maze it does manage to negotiate, rather than the depth into the maze it reaches.

Early in a GP calculation the great majority of programs are very poor, and consequently it may be helpful to use a multiobjective function, which takes into account the performance of the program as measured by several factors. This allows a distinction to be made between programs of modest ability, which are able to complete one or two tasks correctly, and those that are completely hopeless.

GP populations are very much larger than those used in standard GAs. In a conventional GA, the population contains tens or hundreds of strings, but in GP the number of candidate programs may run into several hundred thousand or even millions. It is essential that the population is large, since the fraction of useful programs is small, and the disruption caused by crossover and mutation may be devastating for most programs. The disruption can be reduced by restricting crossover to programs of similar characteristics but any form of alteration to the strings carries with it a high probability of causing a dramatic deterioration in the performance of the string.

No matter what the effect of the genetic operators may be, the evolving programs have a tendency to grow indefinitely unless limits are placed on their size. These limits exist to reduce the quantity of redundant code (known as junk), which otherwise quickly become excessive.

The standard GA procedures can be expanded to include various additional operators, such as encapsulation. This is a particularly valuable step, which collapses a section of code into a single node, thereby protecting it from the disruption which may be brought about by the action of the genetic operators. Encapsulation can be used to create automatically defined functions which are subprograms which may evolve in parallel with the main program. These consist of subroutines or subfunctions which perform specific tasks, and are typically called with one or more variable parameters passed across to them.

Architecture-altering operations are used to extend the scope of GP by providing a means by which branches and arguments within the parse trees may be created or deleted. Permutation, which swaps characters in a tree, and editing, which involves the optimisation of long expressions, are also commonly used.

A potentially serious drawback of GP programs, unless they are very simple, is that they can rarely combine accuracy with lucidity. If it is important that the user appreciate the logic by which the program has reached its conclusion, it may be necessary to accept diminished performance. There are parallels here with the operation of neural networks, which are effective for many types of problem, but, as black boxes, are usually unable to tell us how they reached their conclusions. This lack of transparency may be a serious drawback if we would like to use genetic programs not only to predict or analyse a phenomenon, but also to help us to understand it.

GP has both potential and versatility. It has been applied to a variety of problems in robotic control, image analysis, pattern recognition, music generation and the synthesis of neural networks. Although one might guess that its main value is in the construction of programs for those who cannot write them, it is, at least at its present stage of evolution, of greater value in attempting to solve problems which we do not know how to solve, such as predicting the movement of currencies from the behaviour of the market over the previous week.

7
Conclusion

Although the use of evolutionary computing within science is still in its early stages, scientific applications are already notable for their diversity. Evolutionary methods have been used to optimise the geometry of molecules, the shape of propellers, the properties of polymers and the order in which chemicals are produced in industrial flow lines. They have been used in the study of oil extraction, the natural degradation of toxic chemicals in the environment, spectral deconvolution, the interpretation of microwave spectra and in a wide range of other areas. The chapters that follow provide further illustration of the potential of these intriguing and versatile techniques.

8 Further Reading

1. Back T (1996) Evolutionary Algorithms in theory and practice: evolution strategies, evolutionary programming, genetic algorithms. Oxford University Press, Oxford
2. Bentley PJ, Corne DW (2002) Creative evolutionary systems. Academic, San Diego
3. Cartwright HM, Sztandera LM (2003) Soft computing approaches in chemistry. Springer, Berlin Heidelberg New York
4. Chambers L (ed) (1999) Practical handbook of genetic algorithms. CRC, Boca Raton
5. Clark DE (ed) (2000) Evolutionary algorithms in molecular design. Wiley-VCH, Weinheim
6. Forrest S (ed) (1991) Emergent computation: self-organizing, collective and cooperative phenomena in natural and artificial computing metworks. MIT Press, Cambridge
7. Goldberg DE (1989) Genetic algorithms in search, optimization and machine learning. Addison-Wesley, Reading
8. Holland JH (1975) Adaptation in natural and artificial systems. University of Michigan Press, Ann Arbor
9. Koza JR (1992) Genetic programming: on the programming of computers by means of natural selection. MIT Press, Cambridge
10. Koza JR, Bennett FH III, Andre D, Keane MA (1999) Genetic programming III: Darwinian invention and problem solving. Kaufmann, San Francisco
11. Man KF, Tang KS, Kwong S (1999) Genetic algorithms: concepts and designs. Springer, Berlin Heidelberg New York
12. Michalewicz Z (1996) Genetic algorithms+data structures=evolution programs. Springer, Berlin Heidelberg New York
13. Quagliarella D, Periaux J, Poloni C, Winter G (1998) Genetic algorithms and evolution strategy in engineering and computer science. Wiley, Chichester
14. Russell ST, Norvig P (2003) Artificial intelligence: a modern approach. Prentice Hall, Upper Saddle River
15. Schwefel H-P (1995) Evolution and optimum seeking. Wiley, Chichester
16. http://panizzi.shef. ac.uk/cisrg/links/ea_bib.html

Structure and Bonding, Vol. 110 (2004): 33–53
DOI 10.1007/b13932

Application of Evolutionary Algorithms to Global Cluster Geometry Optimization

Bernd Hartke

Institut für Physikalische Chemie, Christian-Albrechts-Universität, Olshausenstrasse 40, 24098 Kiel, Germany
E-mail: hartke@phc.uni-kiel.de

Abstract This contribution focuses upon the application of evolutionary algorithms to the non-deterministic polynomial hard problem of global cluster geometry optimization. The first years of method development in this area are sketched briefly; followed by a characterization of the current state of the art by an overview of recent application work. Strengths and weaknesses of this approach are highlighted by comparison with alternative methods. Last but not least, current method development trends and desirable future development directions are summarized.

Keywords Global optimization · Atomic clusters · Molecular clusters · Structure · Geometry

1 Introduction 34

2 Historical Development 36

3 Recent Applications 39

3.1 Isolated Atomic Model Systems 39
3.2 Isolated Atomic Main Group Clusters 40
3.3 Isolated Atomic Transition-Metal Clusters 42
3.4 Passivated Clusters 43
3.5 Supported/Adatom Clusters 44
3.6 Isolated Molecular Clusters 44

4 Comparison to Other Methods 45

5 Current and Future Method Development 48

6 References 50

Abbreviations

AM1 Semiempirical Austin method 1
CSA Conformational space annealing
DFT Density functional theory
DFTB Density-functional-based tight binding
EA Evolutionary algorithm

GGA	Generalized gradient approximation (within density functional theory)
HF	Hartree–Fock
LMP2	Local second-order Møller–Plesset perturbation theory
LJ	Lennard-Jones (interparticle potential; used like a chemical symbol for a single atom in this article)
MC	Monte Carlo
MD	Molecular dynamics
NP	Nondeterministic polynomial (problem complexity level)
PES	Potential-energy surface
SAS	imulated annealing
SPC/E	Simple point charge, extended(empirical water potential)
TB	Tight binding
TIP3P	Transferable intermolecular potential with three points (one of Jorgensen's empirical intermolecular water potentials)
TIP4P	Transferable intermolecular potential with four points (another of Jorgensen's empirical intermolecular water potentials)
UHF	Unrestricted Hartree–Fock
n	Number of atoms or molecules in a cluster
m	Population size (number of individuals per generation)

1 Introduction

The modern research area of nanotechnology aims at controlled fabrication and technical use of aggregates of atoms and molecules with typical length scales of nanometers, by making traditional devices smaller and smaller. For a long time, chemistry has been working with small and large molecules, that is, even below the nanometer regime. Some of its subareas, like supramolecular chemistry or cluster chemistry and cluster physics [1–3], now start to progress to larger entities. Therefore, it is only a matter of time until these two research directions "meet" on the nanometer scale.

In their traditional area of single atoms and molecules, theoretical chemists have developed intricate tools that are now able to predict molecular properties with an accuracy rivaling that of corresponding experiments. At the other end of the size scale, theories for the infinitely extended solid state (at least for the periodic case of crystals) are rapidly catching up. Between these domains, however, difficulties remain, and this is the realm of clusters. Structures and properties of medium-sized clusters are neither those that can be extrapolated from the bulk to small scales [4] nor those that could be expected from single particles or the smallest clusters. Instead, one typically finds one or several rather sudden transitions in structures and properties, which still defy explanation [5].

Outside of basic science, clusters also have immediate relevance in many areas, ranging from technical processes like chemical vapor deposition [6] all the way to polar stratospherical clouds in ozone destruction [7]. A direct simulation of the processes relevant in these areas, however, is far beyond current theoretical abilities.

Theoretical calculation of properties of clusters cannot be done without knowledge of the cluster structures. But finding the cluster structure with globally minimal energy turns out to be a nondeterministic polynomial hard problem [8], implying exponential scaling of search space (and hence computational effort) with cluster size. One may argue that the actual cluster structures in experimental or natural situations are not necessarily those of globally minimal energy. But this does not alleviate the problem [5]. Any simulation method has to face this exponential increase of configuration space (unless some strange experimental preparation conditions form the cluster reliably in one known structure – a rare situation). Also, if the experimental structure is not the global minimum, it is a very low energy local one (preferred over the global minimum either for entropic reasons or because it is governed by preparation conditions) – and finding the best low-energy minima within an exponentially growing set of local minima is again best solved by a global minimization approach, which in practice returns not only the global minimum but also a set of low-energy local minima.

As general global optimization tools, evolutionary algorithms (EAs) can be applied to the problem of finding the global minimum-energy structure of atomic and molecular clusters. This idea was first implemented in the early 1990s. In the ensuing years, several research groups contributed to a concerted development effort, adapting general EA tools to the cluster geometry optimization task; pure applications were rare. Shortly before the turn of the millennium, the basic foundations had matured to the degree of turning this idea into an efficient and established procedure. This has caught the attention of more application-oriented research, and thus the past couple of years has seen a rapid increase in the number of pure application papers. This review will briefly sketch the highlights of those development years, characterize the current status by giving an overview of the most recent and current work in this area, for both method development and applications, and outline future directions.

EA applications to global cluster geometry optimization have been reviewed before, see, for example, the general review of EA methods in chemistry by Judson [9] that also discusses some of the early steps towards global cluster geometry optimization. Since then, several other reviews have appeared [5, 10–13], of varying scope and focus. For this reason, and since the literature even in this restricted area starts to outgrow the limits of a single review, the present review does not claim to cover each and every publication, and necessarily it is also biased by the personal views of its author. Also, familiarity with the basic concepts of EAs is assumed. Finally, this is not intended to be a review on theoretical cluster studies in general; therefore, with the exception of Sect. 4, work not using EA methods will be ignored completely.

The remainder of this review is outlined as follows. The historical method development of EA use for global cluster geometry optimization is briefly recalled in Sect. 2. An overview of typical application work in recent years is provided in Sect. 3. In Sect. 4, we take a side glance at other methods to tackle the same and related problems, and briefly discuss advantages and disadvantages of some the prominent alternatives to EAs. Finally, in Sect. 5 recent method development work is summarized, and we try to give some (personal, biased) opinions on which open questions such developments should address in the future.

2 Historical Development

After EAs were invented several times under several different names [14,15] and had already been applied to several different areas in chemistry [9], it was not before 1993 that they started to be used also for solving the problem of global cluster geometry optimization, in a first application to atomic clusters [16] (Si_4), followed by the first application to molecular clusters [17] (benzene dimer, trimer and tetramer). These first steps were made by directly applying the EA in its "pure" form, as advocated at that time. The particle coordinates were encoded as binary strings, standard evolutionary operators (like single-point crossover) were applied to these strings, and there was a constant population size, with child strings replacing parent strings on a generational basis, as well as exponential fitness functions. Interestingly, even at the very start [16] it was emphasized that the choice of a suitable representation (i.e. of a mapping from particle coordinates or configuration space to genetic string space) is important for the success of the method; this was exemplified by different internal coordinate choices resulting in different EA performance.

These first applications were simply too inefficient, and the cluster examples chosen were too small, for them to be serious competition for established methods like simulated annealing (SA) (see Sect. 4); therefore, they went largely unnoticed, and the following year, 1994, did not see much activity in this field at all.

In 1995, several groups struggled with the representation problem. Zeiri [18] switched from binary to real-number encoding of coordinates into genetic strings and introduced various operators for this representation, in an application to Ar_nH_2, $n \leq 12$. Mestres and Scuseria [19] used an adjacency matrix representation, in an application to C_8 and a cluster of n atoms, with a Lennard-Jones, LJ, potential acting between the atoms (LJ_n, $n \leq 13$), in the tight-binding (TB) model (this was attempted again later by Pastor and Poteau [20]). There was even an attempt to circumvent this problem completely, by applying EA methods not to the cluster geometries directly but to the optimization of cluster growth schemes [21].

With hindsight, the most significant publication of that year, however, was a paper by Deaven and Ho [22], introducing several ideas to increase efficiency. They radically cut down population sizes, m, from several dozens to just a handful of individuals; this was compensated for by a departure from EA standards of that time. According to this standard, m individuals were chosen for reproduction, partially weighted by fitness and partially at random; from these $m/2$ pairs crossover and mutation generated m new individuals that constituted the next generation (except for elitist strategies that allowed direct passage of parent individuals into the next generation). In the new Deaven–Ho scheme, all possible unique combinations of parent individuals to pairs were actually realized, and thus $m \times (m-1)$ children were generated. From this intermediate, enlarged pool, m individuals were chosen for the next generation in a sequential fashion, starting with the individual with the lowest energy but then discarding individuals with energies too close to the energies of already-selected individuals. Thus this implementation contained an indirect control over population diversity (via energies, instead of directly via cluster structures), which is now recognized as a key issue in EAs (see Sect. 5).

The second important new ingredient was the radical departure from any kind of string representation, and genetic operators operating on these genetic strings. Instead, Deaven and Ho introduced variants of crossover and mutation that operated directly on the clusters in coordinate space (i.e. on the phenotype rather than on the genotype, as a biologist might say). This idea has two immediate advantages:

1. It makes it much easier to design new and efficient evolutionary operators, and to control their effects and usage, since they operate not on abstract strings but directly in the space were the cluster particles "live".
2. It gets rid of the representation problem by eliminating the need for a representation.

A word of caution is in order here. As natural and elegant as this solution of the representation problem may seem, there is also a downside to it. One may argue that this scheme still uses a representation (a unity representation, so to speak). Every representation of a problem, however, may make finding the solution of the problem easier or more difficult. Therefore, it is conceivable that there are other representations which make the global cluster geometry problem easier than this direct ("unity") representation. In fact, there are indications that this may be true. Researchers studying crystal structure know that seemingly unrelated structures in three-dimensional space are actually related, complementary or even the same in higher-dimensional spaces [23, 24]. So far, however, this observation has attracted only a limited number of followers in the structure optimization community [25].

A third important ingredient in the Deaven–Ho scheme is the use of local optimization to improve each new cluster structure after its formation by the evolutionary operators. Such mixed local/global schemes are called hybrid methods in the EA literature and are rather common there. Here, it took several years before Doye and coworkers [26, 27] demonstrated that this is more than just cosmetics but rather a transformation of a potential-energy surface (PES) difficult for optimization to a simpler one. With this new scheme, Deaven and Ho optimized carbon clusters up to n=60 in a TB model, and managed to find fullerene structures without introducing prior knowledge to these particular geometries.

Although Deaven et al. [28] published another application of the scheme 1 year later (now to the standard benchmark LJ_n, with $n \le 100$), the fundamental importance of that first Deaven–Ho paper was not fully realized immediately; therefore, during the next few years, many groups still largely stuck to the older, more general EA ideas. Also in 1996, Gregurick et al. [29] also incorporated local optimization into their implementation, but still worked with a binary representation; nevertheless they could realize a cluster size scaling of $n^{4.5}$ in the LJ benchmark. They also looked at heterogeneous BAr_n clusters. Shortly after this paper, Niesse and Mayne [30] published a "space-fixed" version of a similar implementation, which was then applied to Si_n, $n \le 10$, on an empirical potential [31]. The same year also saw what probably was the first application of an EA at the density functional theory (DFT) or ab initio level, namely by Tomasulo and Ramakrishna [32], who looked at $(AlP)_n$, $n \leq 6$, directly at the local density approximation-DFT level. The obvious problem of such an approach is the tremendous computational cost, which only allows for small clusters, even if the population size is so small that the reliability of the results can be called into question. The present author proposed to circumvent this problem by using dynamically globally optimized empirical potentials as guiding functions for the search on the ab initio PES [33].

In the following year, 1997, most groups were still investigating effects of various genetic encodings. For example, Zeiri [34] applied his real-number encoding to small argon clusters ($n \leq 10$), while Niesse and Mayne [35] compared various crossover operators in an application to Ar_n and $(H_2O)_n$, $n \leq 13$. To the knowledge of the present author, Pullan [36] was the first to take up the ideas of Deaven and Ho, and investigated their effects on global optimization of LJ clusters [37] (including a direct comparison of genotype and phenotype crossover operators and found the latter to be more favorable), mixed Ar–Xe clusters [38] and benzene clusters [39]; interestingly, in the latter application, he managed to treat nontrivial sizes of molecular clusters, up to n=15. Shortly afterwards, Hobday and Smith [40] also adopted the Deaven–Ho scheme for small and medium-sized carbon clusters (n=3–60), using two different many-body empirical potentials.

The year 1998 was again marked by method development papers, with more of them starting to adopt the ideas of Deaven and Ho. Michaelian [41] advocated a "symbiotic" EA in an application to LJ_n, n=6, 18, 23, 38, 55. In this variant, the optimization of larger clusters is broken down to the optimization of smaller pieces, with the remainder being held fixed. While this is a potentially promising idea to escape bad scaling with cluster size by a more explicit exploitation of the assumed spatial separability of the problem, it remains to be shown that it does not lead the optimization astray in cases of competing structures and structural transitions. Zacharias et al. [42] combined Deaven–Ho-style EAs and SA for the case of Si_n, n=6, 10, 20, using a TB model.

Niesse and Mayne [43] compared binary-coded EA, real-coded EA with local optimization, and basin-hopping, arriving at the conclusion that traditional binary EA without local optimization is not competitive, while the other two are comparable. Interestingly, they argue against the use of phenotype crossover operators. Wolf and Landman [44] explicitly took up the Deaven–Ho algorithm and improved it by twinning mutations and add-and-etch operations, but they failed to treat several LJ cases successfully without using optimal structures of smaller cluster sizes as seeds. The present author managed to show [45] for LJ_n, $n \leq 150$, that seeds were not necessary with a further refined Deaven–Ho algorithm. In this benchmark study, the size scaling of the method was shown to be approximately cubic, opening the way to larger clusters. At the same time, the reintroduction of the well-known EA concept of niches greatly helped to treat the notoriously difficult cases n=38, 75–77, 102–104 without the need to go to significantly longer computation times. These successful developments established the Deaven–Ho variety of EA as a standard tool for global cluster geometry optimization.

Besides method development, papers with focus on applications started to show up, for example, by Michaelian [46] to $(NaCl)_n$, $n \leq 6$, and by Zeiri [47] on Cl and Br ions and atoms in Xe clusters. In a partially successful attempt to rationalize experimental Si cluster mobility data, Ho et al. [48] used an EA with small population size directly on the DFT level, up to a cluster size of n=20. With the previously introduced guiding function method, the present author [49] could use larger populations and obtained agreement with accepted literature structures for DFT calculations on Si clusters up to n=10. In another application of the same technique, water pentamer, hexamer and heptamer clusters were treated successfully at the local second-order Møller–Plesset perturbation theory (LMP2) level, with the additional result of an improved intermolecular water potential [50]. In a benchmark application to transferable intermolecular potential with four points (TIP4P) water clusters [51], known global minimum structures were confirmed up to n=21 and a new structure was proposed for n=22; the still limited size range of this study and its inability to confirm a better size scaling than an exponential one again confirmed the greater difficulty of the molecular cluster problem.

Thus, in retrospect, the period 1998–2000 can be seen as a transition from method development to more routine application work. Recent work of the latter type is summarized in the following section.

3 Recent Applications

Compared to the situation described in earlier reviews, the scope of applications has noticeably widened. In particular, more mixed clusters are now being studied (albeit only binary ones). Also, a few papers have appeared that treat clusters other than bare clusters in a vacuum: hydrogen-passivated silicon clusters and ligand-coated gold clusters have been investigated, as well as clusters on supporting surfaces. Studies of molecular clusters, however, which were attempted at the very beginning of the development described earlier, are still quite rare, and are limited to comparatively small cluster sizes. This bears testimony to the additional difficulty of this task, and perhaps also to the lack of reliable intermolecular potentials.

The application examples summarized in this section are loosely grouped according to the periodic table, for isolated homogeneous and heterogeneous atomic clusters, followed by passivated and supported clusters as well as molecular clusters.

3.1 Isolated Atomic Model Systems

LJ clusters (in LJ units, or as a model for specified rare-gas atom clusters) continue to be used as a benchmark system for verification and tuning in method development. With the work of Romero et al. [52], there are now proposed global minimum structures and energies available on the internet [53], up to n=309. This considerably extends the Cambridge cluster database [54], but the main body of data comes from EA work that used the known LJ lattices (icosahedral, decahedral, and face-centered cubic) as the input. This is obviously dangerous, as exemplified by the recent discovery of the new tetrahedral structural type [55]. One should also note that the discovery of a new structural arrangement, "FD", in Ref. [52] was later refuted [12]. Nevertheless, these data can serve as tight upper limits.

For example, Iwamatsu [56, 57] used the LJ model system as the first test for his algorithm variant; he included an application on mixed Ar–Xe clusters. For the introduction of their "fast annealing EA", Cai and coworkers [58–60] treated LJ_n, $n \leq 116$, confirming the proposed global minimum structures (the present author, however, has difficulties to classify their algorithm as evolutionary; there does not seem to be information exchange between individuals, and populations of solutions are obtained only to maintain some sort of vaguely defined diversity. In fact, in one of their papers [61], the authors dropped the designation "evolutionary" from the name of their algorithm).

While a standardized benchmark is to be favored, one should also note that LJ clusters are not without peculiarities. According to current "lore", many of these clusters are not particularly difficult to optimize globally (e.g., n=55), since the PES is so strongly dominated by the icosahedral structural type and since it is funneled towards the global minimum [62, 63]. At the same time, the notoriously difficult cases n=38, 75–77, 98, 102–104 appear to be very difficult indeed, since in these cases the global minimum is of a different structure that occupies only a tiny fraction of configuration space, separated from the still-dominating icosahedral region by high-energy barriers.

Morse clusters are a potentially more interesting benchmark model system, offering an additional parameter to "tune" the interaction distance. They were used by Roberts et al. [64] to verify the correct functioning of their EA implementation, up to n=50; better performance than by a random search was also demonstrated.

3.2 Isolated Atomic Main Group Clusters

Following in the wake of fullerenes, one of the favorite subjects for cluster studies in the periodic table has been group 4. Pure neutral carbon clusters were the main subject in the seminal paper of Deaven and Ho [22] and in many later studies, for example, by Hobday and Smith [65]. In most cases, the empirical Brenner potential is employed, resulting in small clusters being linear, medium-sized ones occurring as rings, and larger ones as fullerenes. These established findings are not changed by the most recent studies [66].

For carbon, it is of course also tempting to study clusters of clusters, namely aggregation of C_{60} fullerenes [67–69]. This is not really a molecular cluster application since the inner structure of the fullerene, leading to dependence of the particle interaction on relative particle orientation, is largely or completely ignored. The Pacheco–Ramalho empirical potential is used frequently, and fairly large clusters up to n=80 are studied. There appears to be agreement that small fullerene clusters are icosahedral in this model. In contrast to LJ clusters, however, the transition to decahedral clusters appears to occur as early as at n=17; the three-body term of the potential is found to be responsible for this [67].

Mixed clusters of carbon atoms with atoms of other elements have also been investigated. For example, Joswig et al. [70] have looked at small titanium metcars, Ti_mC_n, m=7, 8, n=10–14, at the density functional TB (DFTB) level. They found that the experimentally stablest metcar, Ti_8C_{12}, is energetically not favored.

Silicon is the next heaviest homologue of carbon, but in spite of obvious similarities its chemistry also shows characteristic differences, which is typical for going from the first to the second row of the periodic table. There is a much smaller propensity for π bonding; therefore, it is not surprising that silicon atoms favor quite different cluster structures. There are no linear strings, rings or fullerenes in pure form, although the latter have recently been predicted to be stable with endohedral metal atoms [71, 72], which could even be verified experimentally [73]. Instead, the planar rhombus of Si_4 is the largest planar cluster; all the larger ones have filled three-dimensional forms. Nevertheless, these structures are quite different from corresponding pieces of the bulk crystal. These basic findings are confirmed by various EA applications [49, 65, 74, 75]. With $n \leq 15$, however, the clusters in these studies are not large enough to cover the region of the first structural transition from prolate to near-spherical, which is deduced from drift time measurements of cationic and anionic species [76] to occur near n=25. In contrast to these findings, Wang et al. [77] claimed to see evidence for this transition to occur as early as at n=17, in a generalized gradient approximation (GGA)-DFT study up to n=21. The basic problem here is the lack of a suitably flexible empirical potential. The influence of the three-body term on cluster structures was studied in the papers just cited [74], but so far no empirical potential seems to be able to predict all structures of small silicon clusters qualitatively correctly. Therefore, if further research on more refined functional forms for empirical potentials does not meet with success, one either has to resort to other models, like tensor surface harmonic theory (see. Ref. [78], which also suggests that empirical potentials might perform better for larger clusters), or to brute-force approaches in combination with DFT or ab initio calculations. In the latter case, one faces the problem of extremely large computational cost of the objective function, which is taken up again in Sect. 5.

Clusters of still heavier homologues of carbon offer the prospect of studying the transition to metallic behavior, which is reached with lead in the bulk. This line of study has already been started experimentally [79]. Theory lags behind considerably here, obviously because of the steeply increasing computational effort of first-principles approaches. Nevertheless, a few studies in this direction have started to appear, for example by Wang et al. [80], who looked at Ge_n, $n \leq 21$, within a TB model. Direct comparisons with experiment are still lacking, though.

Examples for the application of EAs to global structure optimization of clusters are not limited to group 4 but now cover most of the periodic table. Lloyd et al. [81] have investigated medium-sized aluminium clusters, n=21–55, with a many-body Murrell–Mottram potential, finding many different structural motifs, including hollow icosahedral geometric shells for the larger clusters. The same group also looked at mixed stoichiometric and nonstoichiometric MgO clusters, with a Coulomb-plus-Born–Mayer potential [82]. There, it turned out that formal charges smaller than 2 may lead to better structures. Pure cadmium clusters were studied by Zhao [83] with an EA at the TB level, plus refinement of the resulting structures with GGA-DFT. Some magic numbers were found and associated with the shell model; otherwise, close-packing dominates, and $n=20$ is found to have bulklike properties. In a combined theoretical and experimental study, Wang et al. [84] found temperature-dependent structural transitions for Sr_n, $n\leq63$, with EA, molecular dynamics (MD) and Monte Carlo (MC) methods on an empirical potential fitted to DFT data.

Metallic clusters were studied by Lai et al. [85] employing Gupta potentials for lead and the alkali metals, and their EA approach was compared with the basin-hopping method. Pastor and Poteau [20] used EA methods on a generic TB level, for smaller clusters up to $n=13$, with an adjacency matrix encoding, tests of several crossover schemes, and seeding with the best structures of smaller $n-1$ clusters.

3.3 Isolated Atomic Transition-Metal Clusters

The group of Zhao is studying a broad spectrum of clusters with a fixed set of methods. They use EA approaches on TB and empirical potentials, sometimes followed by GGA-DFT refinements; electronic and magnetic properties are studied with an spd-band model Hamiltonian in the unrestricted Hartree–Fock (UHF) approximation. Among other systems, they have studied pure clusters of Ag [86], Rh [87], V [88] and Cr [89], and mixed clusters of similar atom types, for example, Co/Cu [90] or V/Rh [91], for cluster sizes up to $n=13$–18.

Of course, there have to be differences in studies of pure and mixed clusters. In the latter, not only the optimal positions of the atoms in the cluster are to be found, but also the optimal distribution of the atom types on these positions. Therefore, additional evolutionary operators may be introduced that directly change this atom-type distribution without affecting the atom positions in space. For the resulting structures, additional questions can be asked: How and how strongly do optimal structures change with cluster composition? Do the different atom types prefer to mix or to segregate within the clusters?

These issues were addressed in mixed-cluster studies for $n\leq56$ with Gupta potentials, for the systems Pd/Pt [92] and Cu/Au [93]. Segregation tendencies were found, and a surprisingly strong dependence of structure on composition, with qualitative changes of cluster structure upon addition of a single different atom.

Not surprisingly, pure gold clusters are attractive objects, and thus have been studied by several groups. For example, Li et al. [94] investigated pure gold clusters with a Gupta potential, while Soler et al. [95] employed DFT and various empirical potentials, using the symbiotic EA of Michaelian [41]. As a common finding, gold clusters appear to have a marked tendency to form disordered or amorphous structures.

Mercury clusters have also been studied with EA methods [96], using an empirical potential as a guiding function for finding global minima on a HF-plus-dispersion potential, for $n\leq15$. This study challenges the usual interpretation of experimental data that locate a transition in bonding type from van der Waals to covalent at $n=13$ and positions it at $n=11$ instead.

This work also highlights the central problem of theoretical treatments of heavy-atom clusters (which is somewhat less pressing for lighter atoms): An appropriately exact first-principles treatment would be coupled-cluster singles and doubles, with perturbative treatment of triples, with large basis sets, including relativistic effects (in particular for gold). This is much too expensive for anything but extremely few atoms (n=2, 3, 4) where global structure optimization is not yet an issue. DFT and empirical potential approaches make global optimizations possible (with markedly different computational costs) but with the cluster structure strongly depending on details of the interparticle interactions, the accuracy of both approaches may not be sufficient. Hence, at present, this is a difficult area for global cluster structure optimization.

3.4 Passivated Clusters

Dangling bonds at the surface make bare clusters highly reactive and hence experimentally difficult to generate and to study. Clusters with surfaces passivated by other elements are more "natural" and easier to handle. They are, however, more difficult to treat theoretically. As with mixed clusters, various different compositions have to be generated and checked, and suitable interparticle potentials for the different species have to be available. Here, accuracy requirements for these potentials may even be greater, since presence and structure of the outer passivation layer may subtly influence the structure preferences of the whole cluster. Therefore, only a few studies on such systems have appeared so far.

Naturally, group 4 elements are again the focus of interest. Structures of hydrocarbons (which may be thought of as partly or completely passivated carbon clusters) have been optimized by Hobday and Smith [65]. Hydrogen-passivated silicon clusters have been studied a few times, for example by Chakraborti and coworkers [97, 98] at the TB level, as well as by Ge and Head [99] at the semiempirical Austin method 1 (AM1) level, with DFT and MP2 refinement calculations; they noted a marked influence of the passivation layer on cluster structures, with $Si_{10}H_{16}$ and $Si_{14}H_{20}$ already exhibiting bulk structure, in stark contrast to bare silicon clusters.

Wilson and Johnston [100] have studied another common case of passivated clusters, namely gold clusters (n=38, 44, 55) protected by an outer layer of thiol ligands. Much larger clusters of this type can be produced routinely in solution, with various types of ligands [101–104]. Wilson and Johnston treated the ligand layer only implicitly, but they could show that for the case of Au_{55} the bare cluster preference of an icosahedral over a cuboctahedral shape is reversed in the presence of a ligand layer. Experimental inference [102] may point in the same direction.

Clearly, much work remains to be done in this subarea. Evolutionary operators need to be refined in order to deal efficiently with clusters and their ligand layers separately, as well as simultaneously. Suitable levels of theory for the interparticle forces have to be found and tested. And, of course, the size gap between application calculations and experiments needs to be closed.

3.5 Supported/Adatom Clusters

Besides bare clusters in a vacuum (cluster beam) and clusters with passivation layers, another important experimental environment for clusters is a (solid) support. Nevertheless, this setup has been addressed in very few EA applications. Zhuang et al. [105] have used the EA method to study surface adatom cluster structures on a metal (111) surface. Miyazaki and Inoue [106] have found that n=13 clusters which are icosahedral in vacuo either form islands or form layered structures upon surface deposition, depending on the substrate–cluster interaction potential.

As in the case of passivated clusters, EA applications in this area have only just begun. Clearly, a solid support substantially changes the optimization problem, simply by its presence (inducing a different symmetry of the surroundings), but also by its (typically periodic) structure. As the first application examples have already shown, the proper choice of interaction potentials will again be crucial. Compared to these major issues, possible local distortions of substrate structures by the presence of adatom clusters can be expected to be a less important effect.

3.6 Isolated Molecular Clusters

One of the first papers on EA applications to global cluster geometry optimization dealt with benzene clusters [17]. This system was used again by Pullan [39] in the development period, for $n \leq 15$. A recent reinvestigation by Cai et al. [107] only went up to n=7, reproducing known results.

Most of the other EA applications to molecular clusters the present author is aware of focus on pure or mixed water clusters. This is not too surprising, considering the facts that water is the most important molecule on this planet and that reliable intermolecular potentials are even harder to produce than reliable interatomic potentials.

For pure water clusters, Qian et al. [108] have used a string-encoded EA in the size region n=2–14, using and comparing the standard empirical water potentials simple point charge/extended (SPC/E; with and without polarization by fluctuating charges), (transferable intermolecular potential with three points (TIP3P) and TIP4P. They found good agreement between TIP3P, TIP4P, SPC/E without polarization and the available experimental information. SPC/E fails for the notorious case of the water hexamer, but it produces good agreement with ab initio calculations for other measures as a function of n, like oxygen–oxygen distances, energies per molecule and average dipole moments. The present author has shown that EA methods are able to perform on a similar level as basin-hopping, for TIP4P clusters up to n=22 [51]. Guimaraes et al. [109] have used water clusters in the size range $11 \leq n \leq 13$ to introduce the new annihilator and history operators. A curiosity of the larger clusters of this type is the conspicuous predominance of structures with all molecules at the surface, which runs counter to chemical intuition that expects at least a single interior molecule at such cluster sizes. With a study of global minimum structures up to n=30 with the highly accurate but computationally very expensive Thole-type method 2, with flexible monomers, potential [110], the present author could show [111] that this behavior is presumably an artifact of the simpler water potentials, and that this question may be answered experimentally by extending IR spectroscopy studies of OH vibrations [112] up to the size n=17.

Water clusters containing simple ions are another area of current experimental and theoretical interest. Accordingly, they are also the subject of EA studies. Chaudhury et al. [113] have used EA methods on empirical potentials to obtain optimized structures of halide ions in water clusters, which they then subjected to AM1 calculations for simulation of spectra. EA applications to alkali cations in TIP4P water clusters [114, 115] have led to explanations of experimental mass-spectroscopic signatures of these systems, in particular the lack of magic numbers for the sodium case and some of the typical magic numbers of the potassium and cesium cases, and the role of dodecahedral clathrate structures in these species.

A common feature of all EA studies of molecular clusters is the limitation to small clusters, compared to atomic clusters. Clearly, this is due to the additional orientational degrees of freedom of the particles in the cluster, or more precisely, due to the intuitively obvious fact that these degrees of freedom are strongly correlated with the positional ones. Nevertheless, EA studies [51] as well as, for example, basin-hopping applications [116] have resorted to dealing with the additional orientational degrees of freedom by introducing certain stages in the algorithm where they are optimized exclusively, with the positional degrees of freedom held fixed. Stepping back a little and allowing for some speculation, this may appear strange, since quite the opposite line of attack seems to be more natural in principle. Since EA methods implicitly rely on a certain minimal degree of separability within the optimization problems to which they are applied (and one may speculate that this is true to some degree for most optimization methods), it appears to be necessary to find a better representation of the molecular cluster problem. If it is possible to build some of the strongest correlations between orientations and positions (which are quite obvious for molecules with highly directional intermolecular bonding, like water molecules) into new "coordinates", the correlations between these coordinates will be smaller, and EA methods (and probably others) will perform better. This is another indication that the Deaven–Ho "phenotype" encoding is perhaps not (yet) the optimal one, at least for molecular clusters.

4 Comparison to Other Methods

Even today there are still cluster structure studies being done without any global optimization tools. It was pointed out several years ago that such approaches can lead to qualitatively wrong results even for very small clusters. In one case of an empirical silicon potential, spurious low-energy planar minima for n=6–8 escaped the attention of researchers using traditional local optimization methods; they were detected only with global optimization techniques [33]. Thus, cluster studies using exclusively local optimization can be taken seriously only if a huge number of minima are generated [117].

The exponentially increasing configuration search space of clusters and the resulting difficulties are also recognized in other areas of theoretical research, for example, in the MD/MC community (although it is usually viewed from a slightly different angle there, under the name of sampling of rare events). For example, several groups noticed that the notorious case of LJ_{38} is not treated adequately even by some of the standard techniques, and thus more advanced sampling techniques had to be established to overcome these problems [118–120]. In spite of this, repeated quenches from standard MD trajectories are sometimes still being employed to find low-energy minimum structures of clusters [121].

Traditional standard SA is also still being used in some cluster studies [122, 123]. There seems to be a hidden consensus in the global cluster geometry optimization community that standard SA is not as efficient as EA methods or basin-hopping, although there appears to be no published "proof" for this. Recently, SA variants incorporating some of the previously mentioned improved sampling techniques have been applied to clusters [124], and there have been many improvements to the basic recipe of SA. So far, however, it is still unclear if any of this improves SA to the point of being competitive with EA methods, for the cluster optimization problem.

Basin-hopping (roughly a combination of MC steps with local minimization, hence its other name MC plus minimization) was first used for protein folding [125]. Wales and Doye [126] applied a variant of this method to the standard LJ cluster benchmark. They could find all the global minima accepted at that time without prior information, up to n=110, including the difficult cases n=38, 75–77, 103–105 (but missing the tetrahedral case n=98), although solving these took at least 1 order of magnitude more computer time than solving the neighboring easier cases. Two years later, this result could be roughly matched by a phenotype EA implementation [45]. In this study, all accepted global minima up to n=150 were located (again missing the tetrahedral type for n=98). Comparing these two algorithms, the EA was slower for smaller clusters but had a better size scaling (n^3 versus approximately n^5), and using niches it could solve the hard cases just mentioned within almost the same time as the neighboring less difficult cases. Thus, within the large error bars of such a rough comparison, these two algorithms were approximately equally "good" at that point in their development (and at that point in history, they were the only ones that produced all known global minima in this size range, without prior information). This is pretty astonishing, considering the totally different way of searching of these two algorithms: in basin-hopping, we have basically only MC steps; whereas in an EA we have mutation operations (thought to be roughly equivalent to MC steps) and crossover operations. The single point of similarity is that both algorithms perform their search not on the actual PES itself, but on a transformed "staircase" surface, since they apply local optimizations at each "step". Ignoring all the details and complications, this apparently enforces one of two possible conclusions: either crossover is not effective at all in EAs (but this runs counter to the practical experiences of most EA practitioners and to the whole EA literature), or the simplification effect induced by the transformation to the staircase surface is so strong that differences in searching this surface do not matter much. The latter view is (implicitly) advocated by Doye and coworkers [26, 27], as already mentioned in Sect. 2.

There is a large group of global optimization methods that explicitly proceed by transforming the PES (unfortunately, the names of these methods often hide this fact, and their similarities). With some stretch of the imagination, optimization by dimensionality increase [25] mentioned in Sect. 2 may also be grouped with the deformation methods. Some of the true deformation methods apply global transformations, like the adiabatic switching method [127]. Others introduce some information of already visited minima into the transformation, like the stochastic tunneling method [128] that "flattens" all barriers above the energy of the lowest minimum found so far. An interesting aspect of this method is that one can even get thermodynamic information from it, after applying a suitable reweighting [129]. Even more information on already visited minima can be introduced, by exponentially biasing the search against revisiting them, as is done in the "energy landscape paving" method of Hansmann and Wille [130]. The end point of this similarity line is reached at algorithms that literally "fill in" visited minima or even transform them into maxima; this could be classified as a tabu search technique that strictly disallows revisiting of minima. An interesting combination of MD and tabu search was recently used [131] for a complete survey of the LJ_{13} PES, improving earlier counts of the total number of minima. Not all of these methods have already been applied to clusters, some only to protein folding. And, again, comparisons to other methods, including EA, are rare or nonexistent. The successes of these methods, however, show that they are apparently capturing an important point.

Again related to tabu search is another line of algorithms that emphasizes a related but slightly different point, namely the need to distribute various search attempts over large areas of configuration space instead of allowing them to happen too close to each other. One example for this type of method is conformational space annealing (CSA), which is based on controlling a suitable distance measure between individuals in configuration space and proceeds by "annealing" this measure, from large distances (corresponding to exploration) to smaller values (corresponding to exploitation of the best area found). CSA was introduced for protein folding [132], but is now also being applied to clusters, and it appears to beat the EA study mentioned earlier [45] for the LJ benchmark [133].

Of course, one may also try to set up cluster structure optimization methods more geared towards clusters from the outset. One obvious candidate is growing larger clusters from smaller ones. This can be done in a rather straightforward manner [134] or in somewhat more involved ways [135]. In any case, the obvious problem with such aufbau methods is sudden size-dependent structural transitions that not only may occur but are speculated to be almost obligatory for clusters [5]. For the same reason, coupling such methods with EA in the form of population seeding, as has often been done, is dangerous. Also, one should not confuse this with the actual cluster growth in experiments, which may proceed quite differently, if recent MD simulations are to be believed [136, 137].

Of course, besides stochastic global optimization methods, there are also many deterministic methods [138–140]. Typical applications of these to clusters have so far been possible only for trivially small clusters, for example, LJ_7 in Ref. [141] or LJ_{13} in Ref. [142]. Clearly, this is no match at all for the stochastic methods that have now reached LJ_{309} in the CSA work of Lee et al. [133].

Big impediments to understanding and properly weighting all the work done in the area of global cluster geometry optimization are the confusing tendency of selling more or less minor variations and/or mixes of existing methods as wholly new methods under new names and the profound scarcity of fair and meaningful method comparisons, in particular across method boundaries, as in Ref. [143], where a convex global underestimator is compared to SA and MC.

5
Current and Future Method Development

With suitable definitions of search functions, EA methods can also be used to locate more features on the PES than just low-energy local and global minima. Chaudhury et al. [144, 145] have implemented methods for finding first-order saddle points and reaction paths, applying them to LJ clusters up to n=30. It remains to be tested, however, if these method can be competitive with deterministic exhaustive searches for critical points for small systems [146], on the one hand, and with the large arsenal of methods for finding saddles and reaction paths between two known minima for larger systems [63], on the other hand.

Cluster aufbau strategies are being refined, for example, by training a neural network with smaller clusters and then using it to supply input structures to an EA treatment [147], for Si_n, n=10, 20. Although this is claimed to give a speedup by a factor of 3–5, the danger of failure at structural transitions is not eliminated. Morris et al. [148] have recently claimed to have found a way to "turn around the building block idea" and learn something about cluster structural trends from the successful building blocks in an EA run. If it is possible to turn this into a working way to generate new low-energy cluster structures, for example, for clusters that have not yet been examined by the EA, this may turn out to be a better way of doing cluster aufbau. But even this strategy may fail at structural transition sizes. Quite apart from reusing successful building blocks to build new clusters, they may help us to gain a better understanding of how clusters prefer to be built in general, and this would be very valuable in itself.

Outside the narrow realm of cluster structures, EA development protagonists have identified "linkage learning" as the decisive ingredient for a really successful and general EA [149, 150]. In a genetic string representation, this is the process of determining which blocks of genes should not be disrupted by crossover operations, in order to arrive quickly at the global optimum. In earlier stages of EA development and application, this problem had to be solved by choosing a suitable combination of problem representation and crossover operator design. Accordingly, this was also an important point in the first phases of applications of EA methods to cluster geometry optimization, see Sect. 2; and the phenotype representation and crossover operation of Deaven and Ho were a decisive step forward in this respect. In linkage-learning, the solution of this problem is left to the EA itself, often on a different level from the actual problem-solving itself, in hierarchical EA schemes. Several problems had to be overcome, for example, it has to be ensured that the actual global optimization does not proceed on a significantly different timescale than the linkage-learning. Successful solutions seem to be available by now, but they have not found their way into the application arena of cluster geometry optimization. In particular, it remains to be investigated if an adaptation of linkage-learning EA methods to cluster geometry optimization should entail a return from phenotype representations and operators back to genotype representations and operators, or if useful linkages can also be learned in the phenotype representation (e.g., between orientations and positions in molecular clusters). Some faint streaks of "learning a little bit of linkage" have already been used several times, under the seemingly different heading of letting the EA itself discover which of a larger set of different crossover operators works best for the problem currently being optimized (see Ref. [151] for a recent example).

Another important design item for future EA method development in the area of cluster geometry optimization may be learned and taken from the comparison with other methods in Sect. 4. There, the aspects of tabulike techniques (not revisiting minima) and keeping a proper distance between individuals in configuration space appeared to be important, as exemplified by energy landscape paving [130] and CSA [133]. It should be noted that loose relatives of these items are present in many EA applications, ranging from minimum-energy difference selection in the original Deaven–Ho implementation [22] to more recent techniques, like niches [45], line-up competition [152] and predators [153] (and, of course, these items were recognized as important in other EA applications, long before the first application of EA to clusters, for example in Ref. [154]). The present author speculates, however, that giving more weight to these aspects may further improve EA performance.

Tabulike techniques and distance between individuals may also become more important for yet another line of development in EA applications to clusters. With empirical potentials, unbiased global optimization of atomic clusters is now possible for several hundred atoms (as was demonstrated for LJ_{309} by CSA [133]; the value of LJ_{250} for a true EA method [12] predates the CSA study by several years, therefore one cannot conclude that CSA beats EA in terms of cluster size). The computational expense of DFT and ab initio methods, however, is several orders of magnitude greater (even if modern linear-scaling methods are used, and even if these methods really do achieve linear size scaling in practice). This strongly limits the possibilities of applying EA methods at these levels of theory. At the same time, there is no doubt whatsoever that this would be a very desirable aim. Even small differences between potential functions may lead to qualitatively different cluster structures, and hence even the small differences between a good empirical potential and a fully correlated ab initio calculation may be detrimental to the results. Fully convincing ways out of this predicament are still lacking. The ways offered so far clearly bear signs of ad hoc solutions. The present author advocates the use of dynamically adapted empirical potentials as guiding functions for the search on the ab initio potentials [33], and this method has had some success for small systems, like Si_n, $n \leq 10$, at the DFT level [49], Hg_n, $n \leq 15$, at a mixed HF–empirical level [96] and $(H_2O)_n$, $n=5, 6, 7$, at the correlated LMP2 level [50]. In this method, it is possible to check how far the empirical potential deviates from the ab initio one at certain points, but this gives only an indirect clue for an answer to the central problem of whether the empirical potential guides the search on the ab initio function towards or even away from the global minimum. And, of course, universal empirical functions that can be made to approximate arbitrary ab initio potentials by suitable parameter-fitting are not (yet) available. Rata et al. [155] propose to avoid this problem by actually doing the global optimization (approximately) at the level of theory desired. The price they have to pay is a radical shrinkage of the EA population to one member, hence the method name "single-parent evolution"; therefore, this method could also be viewed as an MC or basin-hopping search employing EA-style steps. The authors claim that this is more efficient than EA with larger populations for LJ_n, $n \leq 40$ [156] (clusters larger than $n=105$ have apparently not been tried so far). In their applications, however, they could still treat only fairly small clusters, like Si_n, $n \leq 23$ [155] and Fe_n, $n \leq 19$ [156], in spite of operating at the DFTB level, which is even less expensive than DFT, and they seem to need many thousands of generations (15,000 for Fe_{19} in Ref. [156]), which offsets the savings due to the smaller population somewhat. Clearly, more intelligent solutions are needed. One possibility may be to extract information on preferred building blocks [148] from EA calculations on small clusters at a suitable ab initio level, and to use this as local environment preferences in building larger clusters, trying to avoid both traditional empirical functional forms and rigid aufbau schemes.

6 References

1. Haberland H (1994) (ed) Clusters of atoms and molecules. Springer, Berlin Heidelberg New York
2. Martin TP (1996) Phys Rep 273:199
3. Johnston RL (2002) Atomic and molecular clusters. Taylor & Francis, London
4. Jortner J (1992) Z Phys D At Mol Clusters 24:247
5. Hartke B (2002) Angew Chem 114:1534
6. Masi M, Bertani V, Cavallotti C, Carra S (2000) Mater Chem Phys 66:229
7. Zondlo MA, Hudson PK, Prenni AJ, Tolbert MA (2000) Annu Rev Phys Chem 51:473
8. Wille LT, Vennik J (1985) J Phys A 18:L419
9. Judson RS (1997) Rev Comput Chem 10:1
10. Johnston RL, Roberts C (2001) Genetic algorithms for the geometry optimization of clusters and nanoparticles. In: Cartwright H, Sztandera L (eds) Soft computing approaches in chemistry. Springer, Berlin Heidelberg New York
11. Doye JPK (2003) Physical perspectives on the global optimization of atomic clusters. In: Pintér JD (ed) Global optimization – selected case studies. Kluwer, Dordrecht
12. Hartke B (2003) Efficient global geometry optimization of atomic and molecular clusters. In: Pintér JD (ed) Global optimization – selected case studies. Kluwer, Dordrecht
13. Leardi R (2001) J Chemometrics 15:559
14. Fogel DB (1998) (ed) Evolutionary computation: the fossil Record. IEEE, Piscataway
15. Michalewicz Z, Fogel DB (2000) How to solve it: modern heuristics. Springer, Berlin Heidelberg New York
16. Hartke B (1993) J Phys Chem 97:9973
17. Xiao Y, Williams DE (1993) Chem Phys Lett 215:17
18. Zeiri Y (1995) Phys Rev E 51:R2769
19. Mestres J, Scuseria GE (1995) J Comput Chem 16:729
20. Poteau R, Pastor GM (1999) Eur Phys J D 9:235
21. Hartke B (1995) Chem Phys Lett 240:560
22. Deaven DM, Ho KM (1995) Phys Rev Lett 75:288
23. Nesper R, Leoni S (2001) Chem Phys Chem 2:413

24. Kramer P (2001) J Phys A Math Gen 34:1885
25. Faken DB, Voter AF, Freeman DL, Doll JD (1999) J Phys Chem A 103:9521
26. Doye JPK, Wales DJ (1998) Phys Rev Lett 80:1357
27. Doye JPK, Wales DJ, Miller MA (1998) J Chem Phys 109:8143
28. Deaven DM, Tit N, Morris JR, Ho KM (1996) Chem Phys Lett 256:195
29. Gregurick SK, Alexander MH, Hartke B (1996) J Chem Phys 104:2684
30. Niesse JA, Mayne HR (1996) J Chem Phys 105:4700
31. Niesse JA, Mayne HR (1996) Chem Phys Lett 261:576
32. Tomasulo A, Ramakrishna MV (1996) J Chem Phys 105:10449
33. Hartke B (1996) Chem Phys Lett 258:144
34. Zeiri Y (1997) Comput Phys Commun 103:28
35. Niesse JA, Mayne HR (1997) J Comput Chem 18:1233
36. Pullan WJ (1996) PhD thesis. Department of Mathematics and Computing, Central Queensland University
37. Pullan WJ (1997) Comput Phys Commun 107:137
38. Pullan WJ (1997) J Comput Chem 18:1096
39. Pullan WJ (1997) J Chem Inf Comput Sci 37:1189
40. Hobday S, Smith R (1997) J Chem Soc Faraday Trans 93:3919
41. Michealian K (1998) Chem Phys Lett 293:202
42. Zacharias CR, Lemes MR, Dal Pino A Jr (1998) THEOCHEM 430:29
43. White RP, Niesse JA, Mayne HR (1998) J Chem Phys 108:2208
44. Wolf MD, Landman U (1998) J Phys Chem A 102:6129
45. Hartke B (1999) J Comput Chem 20:1752
46. Michaelian K (1998) Am J Phys 66:231
47. Zeiri Y (1998) J Phys Chem A 102:2785
48. Ho K-M, Shvartsburg AA, Pan B, Lu Z-Y, Wang C-Z, Wacker JG, Fye JL, Jarrold MF (1998) Nature 392:582
49. Hartke B (1998) Theor Chem Acc 99:241
50. Hartke B, Schütz M, Werner H-J (1998) Chem Phys 239:561
51. Hartke B (2000) Z Phys Chem 214:1251
52. Romero D, Barron C, Gomez S (1999) Comput Phys Commun 123:87
53. http://www.vcl.uh.edu/~cbarron/LJ_cluster/researchpot.html
54. Wales DJ, Doye JPK, Dullweber A, Hodges MP, Naumkin FY, Calvo F, Hernández-Rojas J, Middleton TF The Cambridge cluster database, http://www-wales.ch.cam.ac.uk/CCD.html
55. Leary RH, Doye JPK (1999) Phys Rev E 60:R6320
56. Iwamatsu M (2001) Comput Phys Commun 142:214
57. Iwamatsu M (2002) Int J Mod Phys C 13:279
58. Cai W, Jiang H, Shao X (2002) J Chem Inf Comput Sci 42:1099
59. Cai W, Shao X (2002) J Comput Chem 23:427
60. Cai W, Yan F, Shao X, Pan Z (2002) THEOCHEM 579:229
61. Jiang H, Cai W, Shao X (2002) Phys Chem Chem Phys 4:4782
62. Doye JPK, Miller MA, Wales DJ (1999) J Chem Phys 111:8417
63. Wales DJ (2003) Energy landscapes. Cambridge University Press, Cambridge
64. Roberts C, Johnston RL, Wilson NT (2000) Theor Chem Acc 104:123
65. Hobday S, Smith R (2000) Mol Simul 25:93
66. Zhang C, Xu X, Wu H, Zhang Q (2002) Chem Phys Lett 364:213
67. Luo Y-H, Zhao J, Qiu S, Wang G (1999) Phys Rev B 59:14903
68. Zhang W, Liu L, Zhuang J, Li Y (2000) Phys Rev B 62:8276
69. Cai W, Feng Y, Shao X, Pan Z (2002) Chem Phys Lett 359:27
70. Joswig J-O, Springborg M, Seifert G (2001) Phys Chem Chem Phys 3:5130.
71. Nagano T, Tsumuraya K, Eguchi H, Singh DJ (2001) Phys Rev B 64:155403
72. Kumar V, Kawazoe Y (2001) Phys Rev Lett 87:045503
73. Hiura H, Miyazaki T, Kanayama T (2001) Phys Rev Lett 86:1733
74. Iwamatsu M (2000) J Chem Phys 112:10976
75. Bazterra VE, Caputo MC, Ferraro MB, Fuentealba P (2002) J Chem Phys 117:11158

76. Hudgins RR, Imai M, Jarrold MF, Dugourd P (1999) J Chem Phys 111:7865
77. Wang J, Wang G, Ding F, Lee H, Shen W, Zhao J (2001) Chem Phys Lett 341:529
78. Wales DJ (1994) Phys Rev A 49:2195
79. Shvartsburg AA, Jarrold MF (1999) Phys Rev A 60:1235
80. Wang J, Zhao J, Ding F, Shen W, Lee H, Wang G (2001) Solid State Commun 117:593
81. Lloyd LD, Johnston RL, Roberts C, Mortimer-Jones TV (2002) Chem Phys Chem 3:408
82. Roberts C, Johnston RL (2001) Phys Chem Chem Phys 3:5024
83. Zhao J (2001) Phys Rev A 64:043204/1
84. Wang GM, Blaisten-Barojas E, Roitberg AE, Martin TP (2001) J Chem Phys 115:3640
85. Lai SK, Hsu PJ, Wu KL, Liu WK, Iwamatsu M (2002) J Chem Phys 117:10715
86. Zhao J, Luo Y, Wang G (2001) Eur Phys J D 14:309
87. Sun H, Ren Y, Luo Y-H, Wang G (2001) Physica B 293:260
88. Sun HQ, Luo Y-H, Zhao JJ, Wang GH (1999) Phys Status Solidi B 215:1127
89. Luo Y, Sun H, Zhao J, Wang G (2000) Commum Theor Phys 34:43
90. Wang J, Wang G, Chen X, Lu W, Zhao J (2002) Phys Rev B 66:014419/1
91. Sun H, Ren Y, Wang G (2001) Phys Status Solidi B 225:301
92. Massen C, Mortimer-Jones TV, Johnston RL (2002) J Chem Soc Dalton Trans 23:4375
93. Darby S, Mortimer-Jones TV, Johnston RL, Roberts C (2002) J Chem Phys 116:1536
94. Li TX, Yin SY, Ji YL, Wang BL, Wang GH, Zhao JJ (2000) Phys Lett A 267:403
95. Soler JM, Beltran MR, Michaelian K, Garzon IL, Ordejon P, Sanchez-Portal D, Artacho E (2000) Phys Rev B 61:5771
96. Hartke B, Flad H-J, Dolg M (2001) Phys Chem Chem Phys 3:5121
97. Chakraborti N, De PS, Prasad R (1999) Z Metallkunde 90:508
98. Prasad R, Chakraborti N (1998) Met Mater Processes 10:203
99. Ge Y, Head JD (2002) J Phys Chem B 106:6997
100. Wilson NT, Johnston RL (2002) Phys Chem Chem Phys 4:4168
101. Schmid G (1992) Chem Rev 92:1709
102. Herrmann M, Kreibig U, Schmid G (1993) Z Phys D At Mol Clusters 26:S1
103. Kiely CJ, Fink J, Brust M, Bethell D, Schiffrin DJ (1998) Nature 396:444
104. Whetten RL, Shafigullin MN, Khoury JT, Schaaff TG, Vezmar I, Alvarez MM, Wilkinson A (1999) Acc Chem Res 32:397
105. Zhuang J, Kojima T, Zhang W, Liu L, Zhao L, Li Y (2002) Phys Rev B 65:045411/1
106. Miyazaki K, Inoue T (2002) Surf Sci 501:93
107. Cai W-S, Yu F, Shao X-G, Pan Z-X (2000) Chin J Chem 18:475
108. Qian J, Stockelmann E, Hentschke R (1999) J Mol Model 5:281
109. Guimaraes FF, Belchior JC, Johnston RL, Roberts C (2002) J Chem Phys 116:8327
110. Burnham CJ, Xantheas SS (2002) J Chem Phys 116:5115
111. Hartke B (2003) Phys Chem Chem Phys 5:275
112. Sadlej J, Buch V, Kazimirski JK, Buck U (1999) J Phys Chem A 103:4933
113. Chaudhury P, Saha R, Bhattacharyya S (2001) Chem Phys 270:277
114. Hartke B, Charvat A, Reich M, Abel B (2002) J Chem Phys 116:2588
115. Schulz F, Hartke B (2002) Chem Phys Chem 3:98
116. Wales DJ, Hodges MP (1998) Chem Phys Lett 286:65
117. Beu TA, Buck U (2001) J Chem Phys 114:7848
118. Munro LJ, Tharrington A, Jordan KD (2002) Comput Phys Commun 145:1
119. Neirotti JP, Calvo F, Freeman DL, Doll JD (2000) J Chem Phys 112:10340,10350
120. Calvo F, Doye JPK (2000) Phys Rev E 63:010902(R)
121. Maillet J-B, Boutin A, Buttefey S, Calvo F, Fuchs AH (1998) J Chem Phys 109:329
122. Akola J, Rytkönen K, Manninen M (2001) Eur Phys J D 16:21
123. Ruette F, Gonzälez C (2002) Chem Phys Lett 359:428
124. Xiang Y, Sun DY, Gong XG (2000) J Phys Chem A 104:2746
125. Li Z, Scheraga HA (1987) Proc Natl Acad Sci USA 84:6611
126. Wales DJ, Doye JPK (1997) J Phys Chem A 101:5111
127. Hunjan JS, Ramaswamy R (2002) Physics/0202032, 11 Feb 2002
128. Wenzel W, Hamacher K (1999) Phys Rev Lett 82:3003

129. Hansmann UHE (2000) Physics/0001028, 13 Jan 2000
130. Hansmann UHE, Wille LT (2002) Physics/0201054, 27 Jan 2002
131. Chekmarev SF (2001) Phys Rev E 64:036703
132. Lee J, Scheraga HA, Rackovsky S (1997) J Comput Chem 18:1222
133. Lee J, Lee I-H, Lee J (2003) Phys Rev Lett 91:080201
134. Koshelev A, Shutovich A, Solov'yov IA, Solov'yov AV, Greiner W (2002) Physics/0207084, 22 Jul 2002
135. Grigoryan VG, Springborg M (2001) Phys Chem Chem Phys 3:5135
136. Ikeshoji T, Hafskjold B, Hashi Y, Kawazoe Y (1996) J Chem Phys 105:5126
137. Ikeshoji T, Torchet G, de Feraudy M-F, Koga K (2001) Phys Rev E 63:031101
138. Horst R, Pardalos PM, Thoai NV (2000) Introduction to global optimization, 2nd edn. Kluwer, Dordrecht
139. Pintér JD (1996) Global optimization in action. Kluwer, Dordrecht
140. Klepeis JL, Schafroth HD, Westerberg KM, Floudas CA (2002) Adv Chem Phys 120:265
141. Maranas CD, Floudas CA (1992) J Chem Phys 97:1992
142. Pintér JD (1999) Extremal energy models and global optimization. In: Laguna M, Gonzalez-Velarde JL (eds) Computing tools for modeling, optimization, and simulation interfaces in computer science and operations research. Kluwer, Dordrecht
143. Foreman KW, Phillips AT, Rosen JB, Dill KA (1999) J Comput Chem 20:1527
144. Chaudhury P, Bhattacharyya SP (1999) Chem Phys 241:313
145. Chaudhury P, Bhattacharyya SP, Quapp W (2000) Chem Phys 253:295
146. Westerberg KM, Floudas CA (1999) J Chem Phys 110:9259
147. (a) Lemes MR, Marim LR, Dal Pino A Jr (2002) Mater Res 5:281; (b) Lemes MR, Marim LR, Dal Pino A Jr (2002) Phys Rev A 66:023203/1
148. Morris JR, Deaven DM, Ho KM, Wang CZ, Pan BC, Wacker JG, Turner DE (1999) In: Davis LD, De Jong K, Vose MD, Whitley LD (eds) IMA volumes in mathematics and its applications, vol 111. Evolutionary algorithms. Springer, Berlin Heidelberg New York
149. Salman AA, Mehrotra K, Mohan CK (2000) Evol Comput 8:341
150. Goldberg DE (2002) The design of innovation – lessons from and for competent genetic algorithms. Kluwer, Dordrecht
151. Kabrede H, Hentschke R (2002) J Phys Chem B 106:10089
152. Yan L (2003) Comput Chem Eng 27:251
153. Manby FR, Johnston RL, Roberts C (1998) MATCH 38:111
154. Wang QH (1987) Biol Cybern 57:95
155. Rata I, Shvartsburg AA, Horoi M, Frauenheim T, Siu KWM, Jackson KA (2000) Phys Rev Lett 85:546
156. Bobadova-Parvanova P, Jackson KA, Srinivas S, Horoi M, Kohler C, Seifert G (2002) J Chem Phys 116:3576

Structure and Bonding, Vol. 110 (2004): 55–94
DOI 10.1007/b13933

Applications of Evolutionary Computation in Structure Determination from Diffraction Data

Kenneth D. M. Harris · Roy L. Johnston · Scott Habershon

University of Birmingham, School of Chemistry, Edgbaston, Birmingham B15 2TT, UK
E-mail: k.d.m.harris@bham.ac.uk

Abstract Evolutionary algorithms are finding increasing use in the study of a wide range of different types of diffraction data. In this chapter, we review recent applications of evolutionary algorithms to study a variety of structural problems. Examples range from the study of disordered materials by analysis of diffuse scattering data to molecular replacement techniques in biological crystallography. However, the main focus of this chapter is on the application of genetic algorithms in the determination of crystal structures directly from powder diffraction data. The principles underlying the application of genetic algorithm techniques in structure determination from powder diffraction data are described, and applications of this methodology to determine molecular crystal structures of considerable complexity are also highlighted. Clearly the opportunity to determine crystal structures directly from powder diffraction data provides a vital alternative to single-crystal X-ray diffraction, when single crystals of appropriate size and quality for single-crystal diffraction cannot be prepared for the material of interest.

Keywords Molecular crystals · Evolutionary algorithms · Powder diffraction · Structure solution · Crystalline solids

1 Introduction 56

2 Background 57

2.1 Crystalline Solids and Crystal Structure 57
2.2 The Relationship Between a Crystal Structure and its Diffraction Pattern 58
2.3 General Aspects of the Opportunity to Use Evolutionary Algorithms in Structure Determination of Crystalline Solids 59

3 Crystal Structure Solution from Powder Diffraction Data Using Genetic Algorithms 60

3.1 Background 60
3.2 Methodology 63
3.2.1 Overview 63
3.2.2 The Structural Fragment 65
3.2.3 The Fitness Function 65
3.2.4 The Mating Procedure 66
3.2.5 The Intermediate Population 67
3.2.6 The Mutation Procedure 68

3.2.7 Natural Selection 69
3.2.8 Lamarckian Evolution 69
3.2.9 The Overall Procedure 70
3.3 Examples of Applications 70
3.3.1 Example 1: The α and β Phases of L-Glutamic Acid 70
3.3.2 Example 2: $Ph_2P(O)\cdot(CH_2)_7\cdot P(O)Ph_2$ 72
3.3.3 Example 3: Structural Rationalization of Oligopeptides 75
3.4 Analysis of the Evolutionary History in Genetic Algorithm Structure Solution Calculations 80
3.5 Combining Energy Information with Powder Diffraction Data in Structure Solution 83
3.6 Application of Parallel Computing Concepts in Genetic Algorithms 87

4 Other Applications of Genetic Algorithms in Crystallographic Areas 88

4.1 Indexing Powder Diffraction Data Using Genetic Algorithms 88
4.2 Biological Crystallography 89
4.3 Analysis of Diffuse Scattering from Disordered Materials 90
4.4 Analysis of Small-Angle X-ray Scattering Data 90
4.5 Miscellaneous Applications 91

5 Concluding Remarks 91

6 References 92

Abbreviations

DE Differential evolution
EDP Evolutionary distribution plot
EPP Evolutionary progress plot
ETP Evolutionary trajectory plot
GA Genetic algorithm
PGA Parallel genetic algorithm

1 Introduction

Single-crystal X-ray diffraction is without question the most important and powerful technique for determining crystal and molecular structures, and applications of this technique led to many of the most important scientific advances that took place during the twentieth century. Scientific landmarks such as the determination of the structure of DNA 50 years ago [1] and determination of the structure of haemoglobin [2] provide potent illustrations of this point. There is every reason to expect that the central importance of single-crystal X-ray diffraction, both in the physical and in the biological sciences, will be sustained throughout the twenty-first century, with particularly exciting opportunities expected to emerge from exploiting the use of modern synchrotron radiation sources.

In part, the widespread and routine use of single-crystal X-ray diffraction techniques in the modern day is a result not only of advances in instrumentation but also of the development of tremendously powerful methods for data analysis, such that, in the vast majority of cases, the crystal structure can be determined almost routinely even from experimental data of only modest quality. Thus, provided a single crystal of sufficient size and quality is available for the material of interest, successful structure determination by analysis of single-crystal X-ray diffraction data is nowadays a virtual formality.

In contrast to this situation that exists for single-crystal X-ray diffraction, analysis of data from many other types of diffraction experiment remains far less routine, and there is therefore considerable scope for the development of new strategies and new techniques for determining structural information from the diffraction data. Among these areas are the determination of crystal structures directly from powder diffraction data (which provides the only route to crystal structure determination when single crystals of appropriate size and quality for single-crystal X-ray diffraction experiments cannot be prepared), the analysis of small-angle scattering data, and studies of diffuse scattering from disordered materials. In recent years there has been growing recognition of the opportunities that evolutionary algorithms may provide within these areas.

In this article, we focus on some of these developments by reviewing recent work on the implementation and application of evolutionary algorithms within crystallographic research, and all of the examples discussed here concern procedures for establishing the best set of structural parameters to fit experimental diffraction data. Although we focus on the case of X-ray diffraction data, there are, in general, equal opportunities to apply the methods discussed here for the analysis of neutron diffraction data.

2 Background

2.1 Crystalline Solids and Crystal Structure

Crystalline solids are characterized by well-defined structures with long-range, three-dimensional periodic ordering of atoms, ions or molecules. Thus, a crystal structure may be represented by specifying the periodicity of the structure (the concept of the "lattice") and the positions of the atoms, ions or molecules within the repeating unit. The repeating unit within the structure is termed the "unit cell". Furthermore, on taking into account the symmetry (described by a three-dimensional space group) of the crystal structure, the unique part of the unit cell contents (the "asymmetric unit") represents only a fraction of the complete unit cell (except in the case of space group $P1$). Thus, the structure of a crystalline solid can be defined in terms of a small number of parameters – the unit cell dimensions $\{a, b, c, \alpha, \beta, \gamma\}$, the space group symmetry S, and the fractional coordinates $\{(x, y, z)_j\}$ of each atom (labelled j) in the asymmetric unit. The process of carrying out structure determination is, in effect, the process of obtaining knowledge of the structural variables $\{a, b, c, \alpha, \beta, \gamma; S; (x, y, z)_j\}$. We note that the electron density distribution, $\rho(\mathbf{r})$, within the unit cell, as discussed in Sect. 2.2, can be approximated directly from knowledge of the structural variables $\{a, b, c, \alpha, \beta, \gamma; S; (x, y, z)_j\}$.

Clearly the structural variables $\{a, b, c, \alpha, \beta, \gamma; S; (x, y, z)_j\}$ provide a description of the bulk structure of the material, and convey no information on structural defects (i.e. local deviations from the perfect periodic structure just discussed) and no information on the structure at the surface of the crystal (which is not necessarily simply a truncated form of the bulk structure, owing to the possible occurrence of surface structural relaxation and/or surface reconstructions).

2.2 The Relationship Between a Crystal Structure and its Diffraction Pattern

In the diffraction pattern from a crystalline solid, the positions of the diffraction maxima depend on the periodicity of the structure (i.e. the dimensions of the unit cell), whereas the relative intensities of the diffraction maxima depend on the distribution of scattering matter (i.e. the atoms, ions or molecules) within the repeating unit. Each diffraction maximum is characterized by a unique set of integers h, k and l (called the Miller indices) and is defined by a scattering vector $\mathbf{h}$ in three-dimensional space, given by $\mathbf{h}=h\mathbf{a}^*+k\mathbf{b}^*+l\mathbf{c}^*$. The three-dimensional space in which the diffraction pattern is measured is called "reciprocal space", whereas the three-dimensional space defining the crystal structure is called "direct space". The basis vectors $\mathbf{a}^*$, $\mathbf{b}^*$ and $\mathbf{c}^*$ are called the reciprocal lattice vectors, and they depend on the crystal structure. A given diffraction maximum $\mathbf{h}$ is completely defined by the structure factor $F(\mathbf{h})$, which has amplitude $|F(\mathbf{h})|$ and phase $\alpha(\mathbf{h})$. In the case of X-ray diffraction, $F(\mathbf{h})$ is related to the electron density $\rho(\mathbf{r})$ within the unit cell by the equation

$$F(\mathbf{h})=|F(\mathbf{h})|\exp[i\alpha(\mathbf{h})]=\int\rho(\mathbf{r})\exp(2\pi i\mathbf{h}\cdot\mathbf{r})\mathrm{d}\mathbf{r}, \tag{1}$$

where $\mathbf{r}$ is the vector $\mathbf{r}=x\mathbf{a}+y\mathbf{b}+z\mathbf{c}$ in direct space ($\mathbf{a}$, $\mathbf{b}$ and $\mathbf{c}$ are the lattice vectors which define the periodicity of the crystal structure). The integration is over all vectors $\mathbf{r}$ in the unit cell. From Eq. (1), it follows that

$$\rho(\mathbf{r})=\frac{1}{V}\sum_{\mathbf{h}}|F(\mathbf{h})|\exp[i\alpha(\mathbf{h})-2\pi i\mathbf{h}\cdot\mathbf{r}], \tag{2}$$

where V denotes the volume of the unit cell, and the summation is over all vectors $\mathbf{h}$ with integer coefficients h, k and l. If the values of both $|F(\mathbf{h})|$ and $\alpha(\mathbf{h})$ could be measured directly from the experimental diffraction pattern, then $\rho(\mathbf{r})$ (i.e. the "crystal structure") could be determined directly from Eq. (2) by summing over the measured diffraction maxima $\mathbf{h}$ [note that this would only be an approximation to $\rho(\mathbf{r})$, as only a finite set of diffraction maxima $\mathbf{h}$ are actually measured experimentally]. However, while the values of $|F(\mathbf{h})|$ can be obtained experimentally [they are related to the measured diffraction intensities $I(\mathbf{h})$], the values of $\alpha(\mathbf{h})$ cannot be determined directly from the experimental diffraction pattern. The lack of experimental information on the phases $\alpha(\mathbf{h})$ constitutes the "phase problem" in crystallography. In order to determine the crystal structure by making use of Eq. (2), it is clearly necessary to make use of methods that provide ways of estimating the phases $\alpha(\mathbf{h})$. The use of such estimated phases $\alpha(\mathbf{h})$ together with the experimentally determined values of $|F(\mathbf{h})|$ in Eq. (2) provides a means of determining, at least approximately, the electron density distribution $\rho(\mathbf{r})$ and hence the crystal structure. Powerful techniques (including direct methods and Patterson methods) have been developed for obtaining estimated phases $\alpha(\mathbf{h})$, and are widely used in the process of carrying out structure determination from single-crystal X-ray diffraction data.

2.3 General Aspects of the Opportunity to Use Evolutionary Algorithms in Structure Determination of Crystalline Solids

In general, the properties of a crystalline solid (except those properties that are influenced by defects or by the crystal surfaces) depend directly on the bulk crystal structure, and may therefore be expressed, at least in principle, as a function of the structural parameters $\{a, b, c, \alpha, \beta, \gamma\}$, S and $\{(x, y, z)_j\}$. As such, there are considerable opportunities to exploit evolutionary algorithms in crystallographic applications based upon exploring the relationships between these structural parameters and the properties of the crystal. For example, if the structure and a particular property are both known experimentally, then evolutionary algorithms could be applied to establish the relationship between the structure and the property in question (e.g. by finding optimum values for the parameters in relationships with known functional forms). Alternatively, if a particular property has been determined either experimentally or computationally for a crystalline solid of unknown structure, and if the dependence of this property on the crystal structure is already known and quantified by a well-defined relationship, then evolutionary algorithms may be applied to establish the best set of structural parameters $\{a, b, c, \alpha, \beta, \gamma; S; (x, y, z)_j\}$ to fit the experimentally or computationally determined property. We focus here on the latter situation, in which the experimental property of interest is the X-ray (or neutron) diffraction pattern for a crystalline solid of unknown structure. The aim is then to determine the set of structural parameters $\{a, b, c, \alpha, \beta, \gamma; S; (x, y, z)_j\}$ that give rise to the best agreement with the experimental diffraction data. As discussed in Sect. 2.2, the functional relationships between crystal structure and diffraction pattern are already known (see also Refs. [3, 4] for a general discussion of diffraction phenomena).

The main focus of this article is the application of genetic algorithm (GA) techniques to solve crystal structures of molecular materials directly from powder diffraction data (Sect. 2). In this application, the aim is to find the set of atomic positions $\{(x, y, z)_j\}$ [knowing the unit cell $\{a, b, c, \alpha, \beta, \gamma\}$ and space group S] that give the optimal representation of an experimental powder diffraction pattern. In such applications, the population in the evolutionary algorithm comprises a set of trial crystal structures, with the "genetic code" for each structure represented by the structural parameters, or an appropriate subset of these parameters. The quality ("fitness") of a given structure is assessed using a crystallographic *R* factor, which quantifies the level of agreement between the experimental diffraction pattern and the diffraction pattern calculated from a given set of structural parameters. We note that the potential energy of the crystal structure may also be computed from the same set of structural parameters, provided an appropriate potential energy function is available for the system of interest. Subsequent sections provide some more details of other applications of evolutionary algorithms within the context of crystallographic applications. General introductions to GAs may be found in Refs. [5–7] and other chapters of this book.

3 Crystal Structure Solution from Powder Diffraction Data Using Genetic Algorithms

3.1 Background

Crystal-structure determination from diffraction data (either for single-crystal or powder samples) can be divided into three stages: (1) unit cell determination ("indexing") and space group assignment, (2) structure solution and (3) structure refinement. In structure solution, the aim is to derive an approximate description of the crystal structure by direct consideration of the experimental diffraction data, but starting from no knowledge of the actual arrangement of atoms, ions or molecules in the unit cell. If the approximate structure solution is a sufficiently good representation of the true structure, a good-quality crystal structure may then be obtained subsequently by refinement of this structural model against the experimental diffraction data. For powder diffraction data, refinement of crystal structures can be carried out fairly routinely using the Rietveld profile refinement technique [8, 9]. In general, structure solution from powder diffraction data is a significantly greater challenge than structure refinement. A schematic overview of the different stages of structure determination from powder diffraction data is shown in Fig. 1.

As many important materials cannot be prepared as single crystals appropriate for conventional single-crystal diffraction studies (nor for synchrotron-based microcrystal diffraction techniques), the ability to solve crystal structures directly from powder diffraction data promises to open up many new avenues of structural science. Although single-crystal and powder diffraction patterns contain essentially the same information, in the former case this information is distributed in three-dimensional space, whereas in the latter case the three-dimensional diffraction data are compressed into one dimension, which generally leads to considerable overlap of peaks in the powder diffraction pattern. As discussed later, such peak overlap gives rise to significant difficulties in solving crystal structures from powder diffraction data. The techniques currently available for structure solution from powder diffraction data can be subdivided into two categories – "traditional" approaches and "direct-space" approaches.

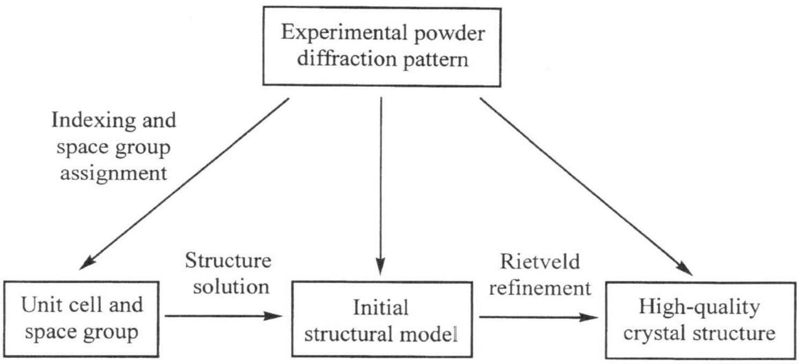

Fig. 1 The different stages involved in determination of a crystal structure from powder diffraction data

The traditional approach [10–16] for solving crystal structures from powder diffraction data is to use the intensities $I(\mathbf{h})$ of individual reflections extracted directly from the powder diffraction pattern. Structure solution may then be attempted by using these $I(\mathbf{h})$ data in the types of structure solution calculation that have been developed for single-crystal diffraction data, such as direct methods or Patterson methods. These methods utilize Eq. (2) in Sect. 2.2 in order to gain an approximate representation of the electron density $\rho(\mathbf{r})$ in the unit cell, using the measured values of $|F(\mathbf{h})|$ [which are derived from the experimentally determined intensities $I(\mathbf{h})$] together with approximate values of the phases $\alpha(\mathbf{h})$ [the methods differ in how they obtain and subsequently improve the estimated values of the phases $\alpha(\mathbf{h})$]. As there is usually extensive overlap of peaks in the powder diffraction pattern (particularly for structures with large unit cells and low symmetry), it is often difficult to extract unambiguous values of the intensities $I(\mathbf{h})$ of the individual diffraction maxima. Unreliable values of the intensities $I(\mathbf{h})$ can lead to severe difficulties in subsequent attempts to solve the structure using such "single-crystal-like" approaches. In order to overcome this problem, we either require improved techniques for extracting and utilizing peak intensities (there have been several developments in this area [17–25]), or we require alternative structure solution strategies (see later) that allow the experimental powder diffraction profile to be used directly in its "raw" digitized form, without any requirement to extract the intensities $I(\mathbf{h})$ of individual diffraction maxima from the experimental powder diffraction pattern.

In the direct-space approach [26–28] for solving crystal structures from powder diffraction data, trial crystal structures are generated in direct space, independently of the experimental powder diffraction data. The powder diffraction pattern for the trial structure is calculated automatically using Eq. (1) in Sect. 2.2 [the structure factor amplitudes $|F(\mathbf{h})|$ obtained using this equation are used to determine the relative intensities $I(\mathbf{h})$ of the diffraction maxima in the powder diffraction pattern]. The suitability of each trial structure is then assessed by direct comparison between the experimental powder diffraction pattern and the powder diffraction pattern calculated for the trial structure. The comparison between the experimental and calculated powder diffraction patterns is quantified using an appropriate R factor. Throughout our work in this field, from the initial demonstration [26] of the direct-space concept to our current work involving the GAs described in Sects. 3.2 and 3.3, we have used the weighted-profile R factor, R_{wp}, which is the R factor normally used in Rietveld refinement [8, 9]. The definition of R_{wp} is

$$R_{wp} = 100 \times \left(\frac{\sum_i w_i \, [y_i(\text{obs}) - y_i(\text{calc})]^2}{\sum_i w_i \, [y_i(\text{obs})]^2} \right)^{1/2}, \tag{3}$$

where y_i(obs) is the intensity of the ith data point in the experimental powder diffraction profile, y_i(calc) is the intensity of the ith data point in the calculated powder diffraction profile and w_i is a weighting factor for the ith data point. Importantly, R_{wp} considers the entire digitized intensity profile, rather than the integrated intensities of individual diffraction maxima; thus, R_{wp} takes peak overlap implicitly into consideration and uses the digitized powder diffraction data directly as measured, without further manipulation. The use of R_{wp} to assess the correctness of a structural model clearly requires that the peak shape and peak width parameters used to construct the calculated powder diffraction pattern are consistent with those that define the experimental powder diffraction pattern. In practice, this can be readily established by prior analysis of the peak shapes and peak widths in the experimental powder diffraction pattern. We note that other definitions of the R factor based on extracted peak intensities may be used to assess the agreement between calculated and experimental powder diffraction data within direct-space structure solution strategies.

The direct-space strategy for structure solution aims to find the trial crystal structure that has the lowest possible R factor, and the approach is equivalent to exploring a hypersurface $R(\mathbf{\Gamma})$ to find the global minimum on the hypersurface. Here $\mathbf{\Gamma}$ represents the set of variables that defines the structure. In principle, any technique for global optimization may be used to locate the lowest point on the $R(\mathbf{\Gamma})$ hypersurface, and much work has been carried out using Monte Carlo [26, 29–35], simulated annealing [36–43] and GA [44–52] search techniques as the basis of direct-space techniques for powder structure solution. In addition, systematic grid search [53–56] and differential evolution (DE) [57, 58] techniques have also been used. In this section, we focus on fundamental and applied aspects of our implementations of GA techniques to achieve global optimization with respect to the $R_{wp}(\mathbf{\Gamma})$ hypersurface.

In all the applications of structure solution discussed here, we assume that the unit cell parameters $\{a, b, c, \alpha, \beta, \gamma\}$ and space group S are already known from prior analysis of the experimental powder diffraction pattern. We also assume that the contents of the unit cell (e.g. the types and number of atoms, ions or molecules) are known, at least to a sufficiently good approximation, but that the positions and structural arrangement of these constituents within the unit cell are not known. The structure is defined in terms of a "structural fragment", which represents an appropriate collection of atoms within the asymmetric unit, and is coded using a set (denoted $\mathbf{\Gamma}$) of variables that represents the positions of the atoms and/or molecules in the unit cell. For a collection of independent atoms, the set $\mathbf{\Gamma}$ would comprise the fractional coordinates $\{(x, y, z)_j\}$ for each of these atoms; however, when the structural fragment comprises a molecule of known constitution, it is greatly advantageous to specify the structural fragment in terms of the position and orientation of the molecule as a whole, together with any variables describing unknown aspects of the intramolecular geometry (such as torsion angles), rather than in terms of the fractional coordinates $\{(x, y, z)_j\}$ of each individual atom. Thus, for a molecular fragment, the position may be defined by the fractional coordinates $\{x, y, z\}$ of the centre of mass or a selected pivot atom, the orientation may be defined by rotation angles $\{\theta, \varphi, \psi\}$ around a set of orthogonal axes passing through the centre of mass or a selected atom, and the intramolecular geometry may be specified by a set of n variable torsion angles $\{\tau_1, \tau_2, ..., \tau_n\}$. These concepts may be extended to the case of two or more (identical or nonidentical) molecular fragments within the asymmetric unit.

In general, the bond lengths, bond angles and any known torsion angles (i.e. if the molecular conformation, or aspects of it, are known a priori) are generally fixed, and may be taken either from standard values for the type of molecule under study or from the known geometry of a similar molecule. Ideally, the structural fragment should include all atoms with significant scattering power (i.e. all non-hydrogen atoms in the case of powder X-ray diffraction) within the asymmetric unit, but it may sometimes be advantageous to omit certain atoms to restrict the number of variables to be optimized (the atoms omitted may be found later by difference Fourier techniques). Clearly the choice of structural fragment in any given structure solution problem is not unique, although certain choices of structural fragment may be significantly advantageous over others.

In this article we focus on fundamental aspects of our implementation of the GA method for structure solution from powder diffraction data (Sect. 3.2), highlighting some examples of the application of this method (Sect. 3.3).

3.2 Methodology

3.2.1 *Overview*

Our GA methodology for structure solution from powder diffraction data is implemented in the program EAGER [44–46, 50–52, 59, 60], and a flow chart describing the operation of this program is shown in Fig. 2. Before running the GA calculation, it is necessary to know the lattice parameters $\{a, b, c, \alpha, \beta, \gamma\}$ and space group S (determined from prior analysis of the powder diffraction pattern) and to make an appropriate choice of the structural fragment. The population comprises a set of trial crystal structures, with each member of the population defined by a set $\mathbf{\Gamma}$ of variables. We note that the values of these variables are real numbers. Each member of the population is characterized uniquely by the values of the variables in $\mathbf{\Gamma}$, which define its genetic code.

The initial population $\mathbf{P}_0$ comprises N_p randomly generated structures. During the GA calculation, the population evolves through a sequence of generations,

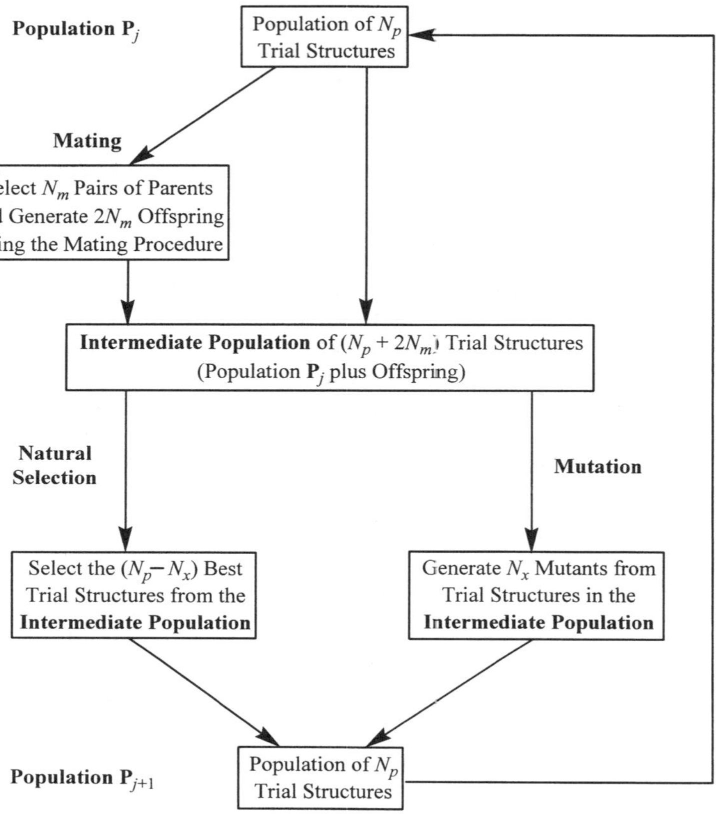

Fig. 2 Flow chart representing the evolution of the population from one generation (population $\mathbf{P}_j$) to the next generation (population $\mathbf{P}_{j+1}$) in the program EAGER

with a given population $\mathbf{P}_{j+1}$ (generation j+1) generated from the previous population $\mathbf{P}_j$ (generation j) by the operations of mating, mutation and natural selection. It is important to note (see later) that mutations create new genetic information within the population, whereas mating serves to redistribute the existing genetic information. The overall scheme for generating population $\mathbf{P}_{j+1}$ from population $\mathbf{P}_j$ in our GA method is summarized in Fig. 2. The number (N_p) of structures in the population remains constant for all generations, and N_m mating operations and N_x mutation operations are involved on passing from population $\mathbf{P}_j$ to population $\mathbf{P}_{j+1}$.

In implementing GA strategies for structure solution, there is considerable scope for variation and flexibility in the methods and rules that may be used to carry out each of the individual evolutionary operations and in the definition of the fitness function. Furthermore, details of the flow chart shown in Fig. 2 may differ from one implementation of the GA to another.

3.2.2
The Structural Fragment

As discussed in Sect. 3.1, each member of the population is a trial crystal structure defined by a set of variables $\mathbf{\Gamma}$, representing the position, orientation and intramolecular geometry of each molecule in the asymmetric unit. The choice of structural fragment for any particular problem is not necessarily unique. For the general case of a rigid molecule, six variables are required: $\mathbf{\Gamma}=\{x, y, z, \theta, \varphi, \psi\}$. For the general case of a molecule with a number, n, of unknown torsion angles, τ_i, each member of the population is defined by $(6+n)$ variables: $\mathbf{\Gamma}=\{x, y, z, \theta, \varphi, \psi, \tau_1, \tau_2, ..., \tau_n\}$.

Clearly, when molecules occupy special positions in the crystal structure, the number of variable degrees of freedom may be reduced from $(6+n)$. For example, when a molecular inversion centre is coincident with a crystallographic inversion centre, the molecule has no translational degrees of freedom, and each member of the population is defined by $(3+n)$ variables: $\mathbf{\Gamma}=\{\theta, \varphi, \psi, \tau_1, \tau_2, ..., \tau_n\}$ (where n is the number of variable torsion angles in half the molecule).

3.2.3
The Fitness Function

The quality of a given trial structure is defined by its fitness, F. The fitness value dictates whether the trial structure survives into subsequent generations (through natural selection) and determines the probability with which it takes part in mating. In our GA, fitness is defined as a function of the weighted-profile R factor R_{wp}. The advantages of using R_{wp} to assess the level of agreement between calculated and experimental powder diffraction patterns were discussed in Sect. 3.1. Calculation of the powder diffraction profile corresponding to any given trial structure requires knowledge of (1) the lattice parameters (to determine peak positions), (2) the atomic positions, as defined by (and obtained directly from) the parameters in the set $\mathbf{\Gamma}$, and atomic displacement parameters (to determine peak intensities), (3) analytical functions to describe the peak shapes and peak widths as a function of diffraction angle 2θ and (4) a description of the background intensity. The shape of a peak in a powder diffraction pattern depends on features of both the instrument and the sample, and different types of peak shape function are appropriate under different circumstances. The most widely used peak shape for powder X-ray diffraction data is the pseudo-Voigt function, which allows flexible variation between Gaussian and Lorentzian character [8,9]. After constructing the calculated powder diffraction pattern for a given trial structure $\mathbf{\Gamma}$, the value of R_{wp} is obtained by fitting the calculated and experimental powder diffraction patterns through variation of the overall scale factor (which serves to put the calculated and experimental powder diffraction patterns on the same absolute intensity scale).

To determine the fitness of a given member of the population from its value of R_{wp}, it is advantageous to consider the following scaled R factor:

$$\rho = \frac{R_{wp} - R_{min}}{R_{max} - R_{min}}, \tag{4}$$

where R_{min} and R_{max} are the lowest and highest values of R_{wp} in the population, respectively. We note that the value of ρ lies in the range $0 \le \rho \le 1$. The fitness is then expressed as an appropriate function of ρ. Examples of fitness functions that have been used in our work are

$$\text{exponential:}\quad F(\rho) = \exp(-s\rho), \tag{5}$$

$$\text{tanh:}\quad F(\rho) = {}^{1}/_{2}\{1-\tanh[2\pi\,(2\rho-1)]\}, \tag{6}$$

$$\text{power:}\quad F(\rho) = 1-\rho^{n}, \tag{7}$$

and

$$\text{cosine:}\quad F(\rho) = {}^{1}/_{2}\,[1+\cos(\pi\rho/2)]. \tag{8}$$

In each case, $F(\rho)$ takes its highest value [i.e. $F(\rho)=1$] when $\rho=0$ (i.e. $R_{wp}=R_{min}$) and takes its lowest value when $\rho=1$ (i.e. $R_{wp}=R_{max}$). The values of R_{min} and R_{max} are continually updated as the population evolves during the GA calculation, representing dynamic scaling of the fitness function.

3.2.4
The Mating Procedure

The probability of selecting a given structure to take part as a parent in mating is related to its fitness, with structures of high fitness more likely to be selected. In our procedure for selecting parents, a structure (with fitness F) is chosen from the population at random and a random number κ (with $0 \le \kappa \le 1$) is generated. The selected structure is then allowed to take part in mating if $F > \kappa$. This selection procedure is continued in order to find a second structure that is allowed to mate with the first structure. Pairs of structures selected consecutively in this way are allowed to mate with each other, until the required number (N_m) of mating operations has been carried out. Note that a given structure could be selected several times for mating in a given generation.

We now consider some specific methods that may be used to generate offspring by combining the variables in the sets Γ for the two selected parents. For a rigid molecule defined by six variables $\{x, y, z, \theta, \varphi, \psi\}$, one approach for mating is single-point crossover in which the sets of variables defining the two selected parents are cut and spliced between the positional and orientational variables to produce two offspring. Thus, the parents $\{x_a, y_a, z_a | \theta_a, \varphi_a, \psi_a\}$ and $\{x_b, y_b, z_b | \theta_b, \varphi_b, \psi_b\}$ would lead to the two offspring $\{x_a, y_a, z_a | \theta_b, \varphi_b, \psi_b\}$ and $\{x_b, y_b, z_b | \theta_a, \varphi_a, \psi_a\}$. Such single-point crossover at a fixed point has the potential disadvantage that, for a given pair of parents, it will always lead to the same pair of offspring, and might therefore contribute to loss of diversity within the population. Although the use of single-point crossover between the positional and orientational variables is attractive in view of the physical significance associated with separating the positional and orientational information, there is no guarantee that this represents the most efficient approach for finding the optimal structure solution, and alternative approaches, such as single-point crossover at randomly selected positions within the set of variables or multiple-point crossover, can have several advantages. In these methods, a given pair of parents could produce several different possible pairs of offspring. An alternative procedure is to take the six variables from each parent and distribute them (on a random basis) between the two offspring, with no restriction on which combination of variables may come from each parent.

For a molecule with two variable torsion angles, one method for mating divides the eight variables that define each parent into four groups $\{x, y, z|\theta, \varphi, \psi|\tau_1|\tau_2\}$. For mating between two selected parents, the four groups are divided into two sets of two groups, which can be done in three different ways:

1. $\{x, y, z|\theta, \varphi, \psi\}$ and $\{\tau_1|\tau_2\}$.
2. $\{x, y, z|\tau_1\}$ and $\{\theta, \varphi, \psi|\tau_2\}$.
3. $\{x, y, z|\tau_2\}$ and $\{\theta, \varphi, \psi|\tau_1\}$.

In a given mating operation, one of these ways of dividing the four groups is chosen (with equal probability), and two offspring are generated by taking the first set of two groups from one parent and the second set of two groups from the other parent, and vice versa. Thus, mating the parents $\{x_a, y_a, z_a|\theta_a, \varphi_a, \psi_a|\tau_{1a}|\tau_{2a}\}$ and $\{x_b, y_b, z_b|\theta_b, \varphi_b, \psi_b|\tau_{1b}|\tau_{2b}\}$ will lead with equal probability to one of the following pairs of offspring:

1. $\{x_a, y_a, z_a|\theta_a, \varphi_a, \psi_a|\tau_{1b}|\tau_{2b}\}$ and $\{x_b, y_b, z_b|\theta_b, \varphi_b, \psi_b|\tau_{1a}|\tau_{2a}\}$.
2. $\{x_a, y_a, z_a|\theta_b, \varphi_b, \psi_b|\tau_{1a}|\tau_{2b}\}$ and $\{x_b, y_b, z_b|\theta_a, \varphi_a, \psi_a|\tau_{1b}|\tau_{2a}\}$.
3. $\{x_a, y_a, z_a|\theta_b, \varphi_b, \psi_b|\tau_{1b}|\tau_{2a}\}$ and $\{x_b, y_b, z_b|\theta_a, \varphi_a, \psi_a|\tau_{1a}|\tau_{2b}\}$.

Clearly, many different options exist for mating, each of which may have greater or lesser advantages in different circumstances. For systems involving larger numbers of variables than the examples given here, more complex rules governing the mating procedure may be adopted.

It is important to recognize that the mating operation generates new structures by redistributing the existing genetic information in different ways, but does not actually create any new values of the individual genetic variables. New values of the genetic variables are instead introduced into the population by the mutation procedure, discussed in Sect. 3.2.6.

3.2.5
The Intermediate Population

As the number of mating operations in each generation is N_m and each mating operation leads to two offspring, the total number of offspring produced in each generation is $2N_m$. An intermediate population ($\mathbf{I}_{j+1}$) containing N_p+2N_m structures is then constructed, taking the N_p structures from the previous generation ($\mathbf{P}_j$) and the $2N_m$ offspring generated by the mating procedure. At this stage the values of R_{min} and R_{max} for the intermediate population are determined, and the values of fitness for all members of the intermediate population are recalculated. Although each structure carried through to the intermediate population from the previous generation has the same value of R_{wp} as in the previous generation, the value of the fitness may change (as the values of R_{min} and R_{max} may differ between populations $\mathbf{P}_j$ and $\mathbf{I}_{j+1}$). If two or more structures are identical within predefined tolerance limits, all but one of these structures is eliminated from the intermediate population. The structures in the intermediate population are then ranked according to their fitness, in preparation for natural selection.

3.2.6
The Mutation Procedure

In each generation, a number (N_x) of mutant structures are generated in order to introduce new genetic information within the population, and thus to help to maintain diversity. In our mutation procedure, N_x parent structures are selected at random from the intermediate population, and a new mutant structure is generated from each selected parent by introducing random changes to the values of one or more variables in its genetic code ($\mathbf{\Gamma}$). The changes that are made to selected variables in generating the mutants may either be new random values (static mutation) or small random displacements from the existing values (dynamic mutation). It is important to note that the parent structures used to create the mutants are not replaced by the mutants, but remain within the intermediate population.

As an example, for a molecule with two torsional degrees of freedom, one method for carrying out mutation is to randomly select two of the four groups of variables $\{x, y, z|\theta, \varphi, \psi|\tau_1|\tau_2\}$ and to assign a new random value to one variable within each of the selected groups. For systems involving greater numbers of variables, more complex rules governing the mutation procedure may be adopted.

In principle, the mutation procedure could be introduced in several different ways within the overall scheme for converting population $\mathbf{P}_j$ to population $\mathbf{P}_{j+1}$. However, it is important that the mutant structures are allowed the opportunity to take part in mating operations before they are subjected to the process of natural selection (in the scheme shown in Fig. 2, the mutant structures produced in a given generation pass directly into the next generation, and they are allowed to participate in mating events in that generation before they are considered in the natural selection process). Thus, while many of the mutants will themselves represent poor quality structures (and will be rejected subsequently from the population at the natural selection stage), they may nevertheless contain useful genetic information which may be passed into the population through the mating procedure. Indeed, in our GA implementation, we actively encourage mutant structures to take part in mating events by giving them an artificially high fitness value in the generation immediately after they are produced.

In dynamic mutation, the selected variables are subjected to random displacements from their values in the parent structure. Thus, for a particular variable ξ in the set $\mathbf{\Gamma}$, the new (mutated) value ξ_m is given by

$$\xi_m = \xi_p + (\kappa \times \Delta\xi_{max}), \tag{9}$$

where ξ_p is the value of ξ in the parent structure, κ is a random number between -1 and +1, and $\Delta\xi_{max}$ is the maximum allowed displacement. Dynamic mutation can be particularly useful for fine-tuning the population in the later stages of the GA calculation. Clearly there is considerable scope for optimizing the mutation strategy, for example, by using static mutation in the initial stages and then introducing dynamic mutation in the later stages of the GA calculation.

3.2.7
Natural Selection

In the natural selection procedure, the N_p–N_x best (highest fitness) members of the intermediate population $\mathbf{I}_{j+1}$ are selected. These structures are taken together with the N_x mutant structures generated from members of the intermediate population $\mathbf{I}_{j+1}$ in order to construct the new population $\mathbf{P}_{j+1}$ (which comprises N_p structures). The new values of R_{min} and R_{max} for the new population $\mathbf{P}_{j+1}$ are evaluated, and the fitness of each structure is then determined.

One consequence of the natural selection procedure is that the value of R_{wp} for the best structure in population $\mathbf{P}_{j+1}$ must be less than or equal to the value of R_{wp} for the best structure in population $\mathbf{P}_j$, and thus R_{min} must either decrease or remain unaltered on moving from one generation to the next generation. The population size (N_p) remains constant from one generation to the next generation, and the best structures in a given generation are almost certain to be carried forward into the next generation (i.e. it is likely that the structures passing through the natural selection process will comprise the fittest members of the population in the previous generation together with some of the offspring produced from this generation). The overall quality of the population, assessed from the average value of R_{wp} (denoted R_{ave}), generally improves from one generation to the next generation. However, as mutants are included in the calculation of R_{ave}, the value of R_{ave} may sometimes increase slightly on passing from one generation to the next generation.

3.2.8
Lamarckian Evolution

In a recent implementation of our GA method [60], each structure generated during the GA calculation is subjected to local minimization of R_{wp} with respect to the structural variables in the set $\mathbf{\Gamma}$, and only these minimized structures are used subsequently in the GA calculation. Importantly, the introduction of local minimization in this way has been found to improve the reliability and reproducibility of finding the correct structure solution (e.g. in repeated runs from different random initial populations) and to lead to a reduction (by at least a factor of 10) in the number of generations required to find the correct structure solution. These advantages of introducing local minimization of R_{wp} within the GA calculation may be attributed to a favourable combination of stochastic (i.e. the GA) and deterministic (i.e. the minimization) components within the global optimization strategy. As the genetic characteristics of each structure sampled on the R_{wp} hypersurface become modified depending on the nature of its local environment in the minimization step, the GA method incorporating local minimization represents Lamarckian (rather than Darwinian) evolution.

In view of the advantages just described, the Lamarckian GA method incorporating local minimization of R_{wp} is now our standard approach for tackling structure solution from powder diffraction data.

3.2.9
The Overall Procedure

The structure solution calculation is started from a randomly generated initial population, and the complete evolutionary cycle involving mating, mutation and natural selection is repeated for a specified number (N_g) of generations, or until some predetermined termination criterion is satisfied (such as reaching a sufficiently low value of R_{min}). As the GA is a stochastic procedure, there is no absolute guarantee that the global minimum will actually be located in a given calculation with a finite number of generations. Thus, a good strategy is to repeat the calculation several times from different random initial populations – finding the same structure or very similar structures repeatedly is a strong indication that these structures represent the global minimum.

3.3
Examples of Applications

Several structures of varying degrees of complexity have been solved from powder diffraction data using our GA method. As illustrative examples, three case studies are highlighted. First, as a demonstration of the success of the method, we describe (Sect. 3.3.1) structure solution calculations for two previously known crystal structures – the α and β polymorphs of L-glutamic acid. The other examples (Sects. 3.3.2, 3.3.3) describe structure solution calculations in cases for which the structures were unknown beforehand, thus representing real examples of the application of the methodology in carrying out complete structure determination from powder X-ray diffraction data. In all cases, the powder X-ray diffraction data were recorded using a standard laboratory powder X-ray diffractometer (Siemens D5000; transmission; Ge monochromated, Cu $K_{\alpha 1}$; linear position sensitive detector, covering 8° in 2θ).

3.3.1
Example 1: The α and β Phases of L-Glutamic Acid

L-Glutamic acid, $HO_2C(CH_2)_2CH(NH_2)CO_2H$, is known to exist in two different polymorphic forms, denoted the α and β phases. In both crystal structures [61, 62], the L-glutamic acid molecules are in the zwitterionic form (Fig. 3a). Both structures have the orthorhombic space group $P2_12_12_1$. The unit cell parameters for the α phase are a=10.28 Å, b=8.78 Å, c=7.07 Å. The unit cell parameters for the β phase are a=5.16 Å, b=17.30 Å, c=6.95 Å. Both structures have one molecule in the asymmetric unit.

In our GA structure solution calculations for both polymorphs, the structural fragment comprised all non-hydrogen atoms of the L-glutamic acid molecule (Fig. 3b). Standard geometries (bond lengths and bond angles) were used, with

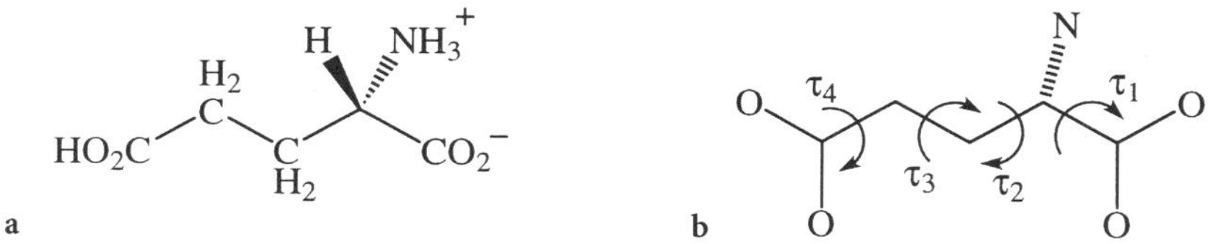

Fig. 3 a Molecular structure of L-glutamic acid in the zwitterion form. **b** Structural fragment used in the genetic algorithm structure solution calculations for L-glutamic acid, showing the variable torsion angles

all C–O bond lengths taken to be equal (the C–O single and C=O double bonds could be assigned subsequently during Rietveld refinement). The four torsion angles $\{\tau_1, ..., \tau_4\}$ defining the conformation of the L-glutamic acid molecule are indicated in Fig. 3b. The position of the structural fragment was defined by the $\{x, y, z\}$ coordinates of the central carbon atom of the molecule.

The GA calculation involved the evolution of 100 generations of a population of 100 structures. In each generation, 200 offspring (100 pairs of parents) and ten mutations were generated. For mating and mutation, the ten variables were considered in terms of six groups $\{x, y, z|\theta, \varphi, \psi|\tau_1|\tau_2|\tau_3|\tau_4\}$. To carry out a mating operation between two parents, the six groups from each parent were distributed between the two offspring, with no restriction on which combination of groups may come from each parent (in each mating operation, the combination of groups coming from each parent was determined on a random basis). In carrying out the mutation procedure on a selected structure, two groups were selected at random, and a random change was made to one variable within each of the selected groups.

The progress of the GA structure solution calculation is assessed from the Evolutionary Progress Plot (EPP), which shows the best (R_{min}) and average (R_{ave}) values of R_{wp} for the population as a function of generation number. The EPPs for the α and β phases are shown in Fig. 4, from which it is clear that the GA structure solution calculation converges rapidly in both cases.

The best structure solution (i.e. the structure with lowest R_{wp} in the final generation) for the α phase is shown in Fig. 5a and the best structure solution for the β phase is shown in Fig. 5b. For comparison, the known crystal structures [61, 62] of the α and β phases are also shown. In each case, the structure solution obtained from the GA calculation is in excellent agreement with the known structure. In each case, the maximum distance between an atom in the structure solution and the corresponding atom in the known crystal structure is less than 0.5 Å, and the structure refines readily (using the Rietveld method) to the known crystal structure. We emphasize that the L-glutamic acid molecule has a significantly different conformation in the α and β phases, and the GA structure solution calculations have successfully found the correct conformation for each phase.

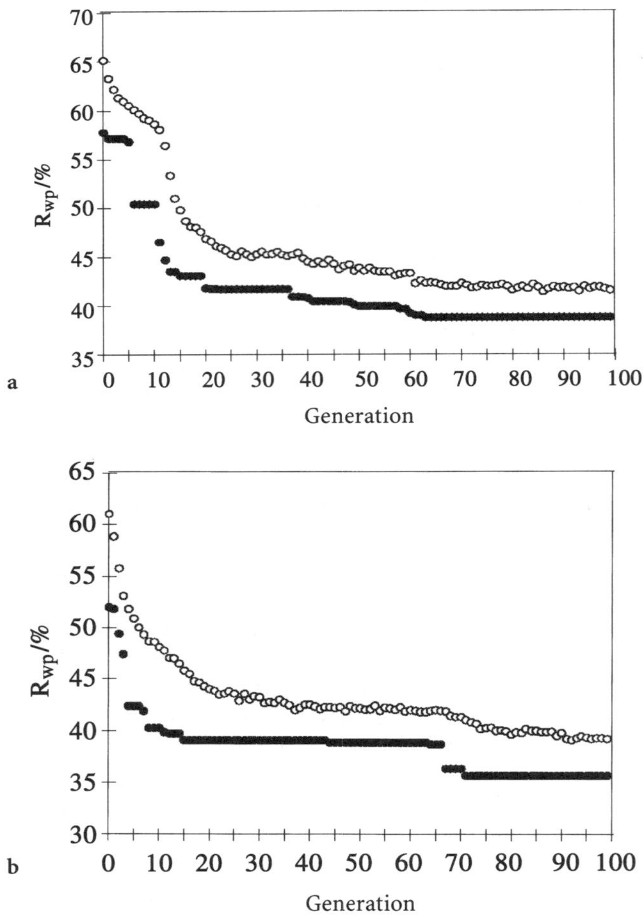

Fig. 4 Evolutionary Progress Plots showing the evolution of R_{ave} (*open circles*) and R_{min} (*filled circles*), as a function of generation number, in the GA structure solution calculations for **a** the α phase and **b** the β phase of L-glutamic acid

3.3.2
Example 2: $Ph_2P(O)\cdot(CH_2)_7\cdot P(O)Ph_2$

The structure determination of $Ph_2P(O)\cdot(CH_2)_7\cdot P(O)Ph_2$ [50] represents one of the most complex molecular crystal structures that has been solved directly from powder diffraction data. Structure solution of this previously unknown structure was carried out using our GA method, involving 18 degrees of freedom with 12 variable torsion angles.

The powder X-ray diffraction pattern of $Ph_2P(O)\cdot(CH_2)_7\cdot P(O)Ph_2$ was indexed using the program ITO [63], giving the unit cell a=12.59 Å, b=10.20 Å, c=22.89 Å, β=105.5°. From systematic absences, the space group was assigned

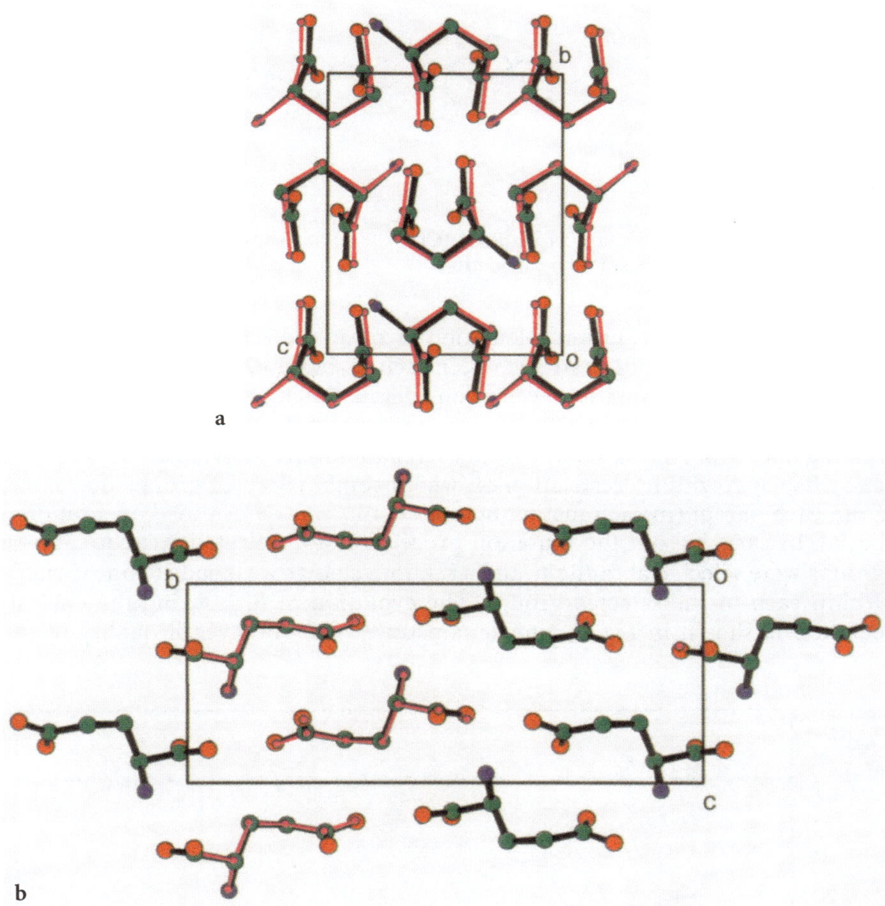

Fig. 5 Comparison between the position of the structural fragment in the best structure solution obtained in the GA structure solution calculation (*lilac*) and the positions of the corresponding atoms in the known crystal structure (*black*) for **a** the α phase and **b** the β phase of L-glutamic acid

as $P2_1/n$, and density considerations suggested that there is one molecule in the asymmetric unit.

In the GA structure solution calculation, the structural fragment comprised all non-hydrogen atoms of the $Ph_2P(O)\cdot(CH_2)_7\cdot P(O)Ph_2$ molecule. Standard bond lengths and bond angles were used, and the atoms of each phenyl ring and all atoms directly bonded to the ring were constrained to be coplanar. The molecule was subjected to translation and reorientation within the unit cell, together with variation of all 12 torsion angles (Fig. 6) that define the molecular conformation. Thus, each structure in the GA calculation was defined by 18 variables $\{x, y, z, \theta, \varphi, \psi, \tau_1, \tau_2, \tau_3, \ldots, \tau_{12}\}$.

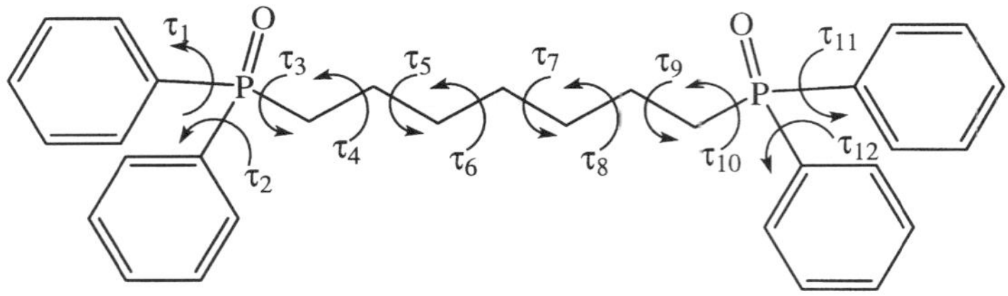

Fig. 6 The molecular structure of $Ph_2P(O){\cdot}(CH_2)_7{\cdot}P(O)Ph_2$, showing the variable torsion angles in the GA structure solution calculation

The GA structure solution calculation involved the evolution of 500 generations of a population of 100 structures. In each generation, 100 offspring (50 pairs of parents) and 20 mutations were considered. For mating and mutation, the 18 variables were subdivided into 14 groups $\{x, y, z|\theta, \varphi, \psi|\tau_1|\tau_2|\tau_3|...|\tau_{12}\}$. In the mating operation, the 14 groups from each parent were distributed between the two offspring, with no restriction on which combination of groups may come from each parent (in each mating operation, this was determined on a random basis). In carrying out the mutation procedure on a selected structure, seven groups were selected at random, and a random change was made to one variable within each of the selected groups. The evolution of R_{wp} during the GA calculation is shown in Fig. 7, and demonstrates that the overall quality of the

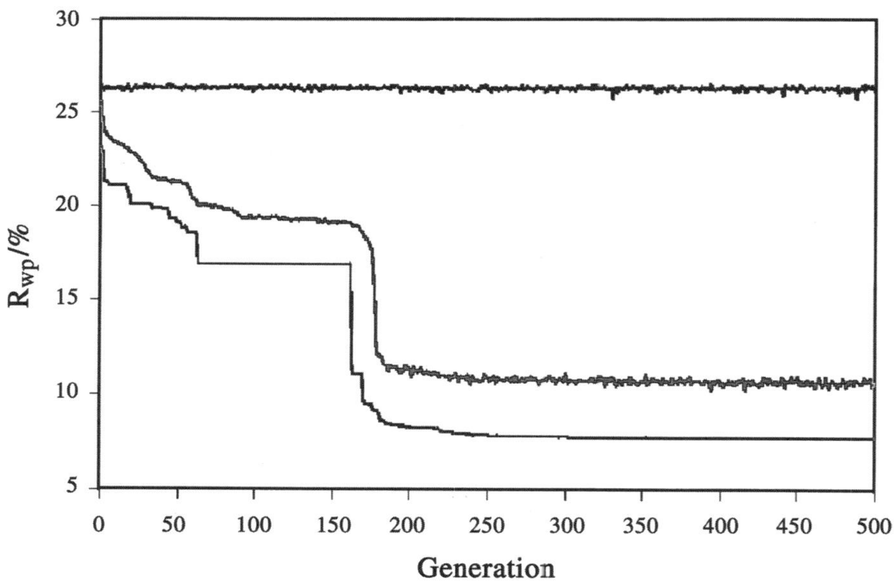

Fig. 7 Evolutionary Progress Plot showing the evolution of R_{wp} for the best structure in the population (R_{min}; lower line), the average value of R_{wp} for the population (R_{ave}; middle line) and the highest value of R_{wp} in the population (R_{max}; upper line) as a function of generation number in the GA structure solution calculation for $Ph_2P(O){\cdot}(CH_2)_7{\cdot}P(O)Ph_2$

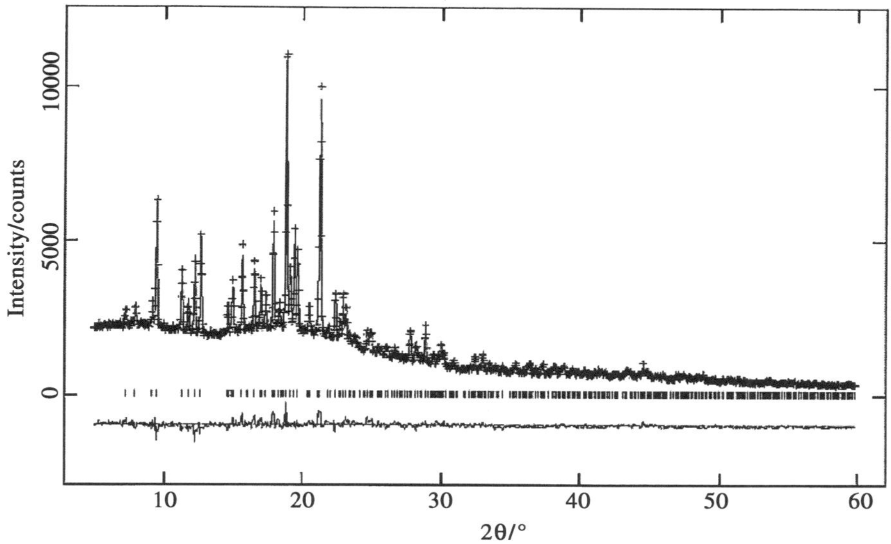

Fig. 8 Experimental (+), calculated (*solid line*) and difference (*lower line*) powder X-ray diffraction profiles for the Rietveld refinement of $Ph_2P(O).(CH_2)_7.P(O)Ph_2$. Reflection positions are marked. The calculated powder diffraction profile is for the final refined crystal structure

population improves as the population evolves. The lowest value of R_{wp} in the population decreases significantly, with evidence for a particularly significant evolutionary event at generation 163.

The structure with lowest R_{wp} in the final generation was taken as the starting model for Rietveld refinement. The positions of all non-hydrogen atoms were refined, with standard geometric restraints applied to bond lengths and bond angles. The final Rietveld refinement (Fig. 8) gave R_{wp}=5.0% and R_p=3.8%. In the crystal structure (Fig. 9), the molecule adopts an unexpected (but completely plausible) conformation, with one gauche bond in the $(CH_2)_7$ chain and the other parts of the chain close to all-trans conformations.

3.3.3
Example 3: Structural Rationalization of Oligopeptides

Knowledge of the conformational properties and interactions in model oligopeptide systems can yield important insights concerning the structural properties of polypeptide sequences in proteins. In many cases, however, the target materials cannot be prepared in the form of single crystals appropriate for single-crystal X-ray diffraction studies, and in such cases structure determination from powder diffraction data represents the only viable route towards structural understanding and rationalization. The Lamarckian GA was used in the structure solution calculations in all the cases discussed below.

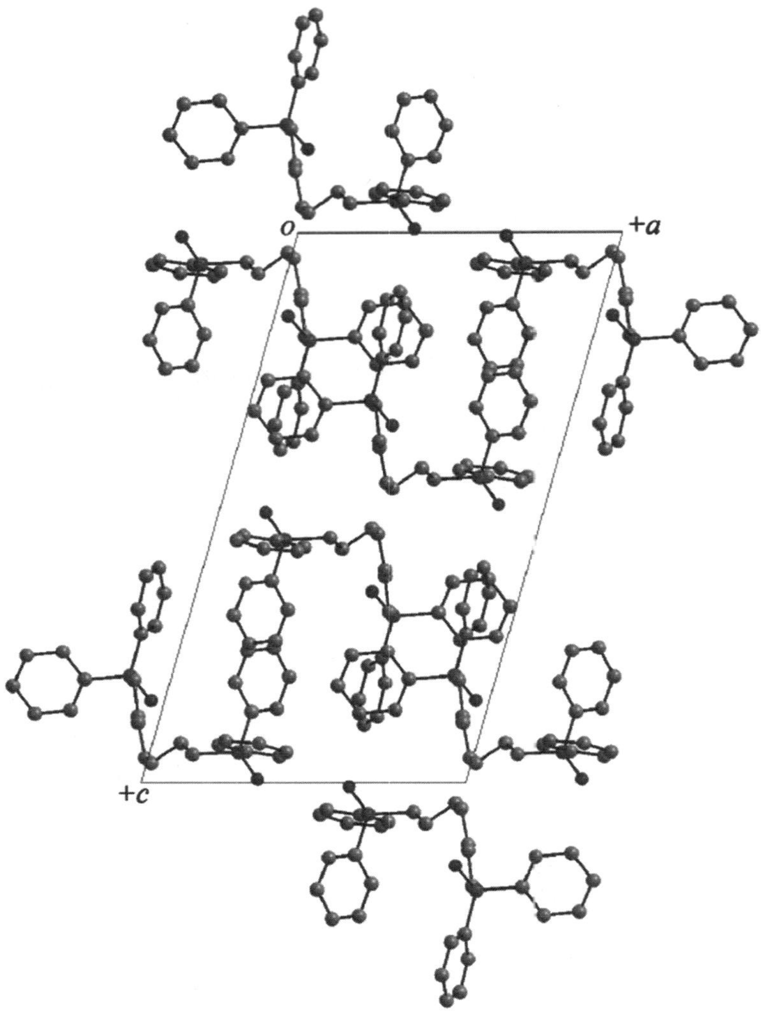

Fig. 9 Final refined crystal structure of $Ph_2P(O)\cdot(CH_2)_7\cdot P(O)Ph_2$ (hydrogen atoms not shown) viewed along the *b*-axis

The first example concerns the oligopeptide Phe–Gly–Gly–Phe [64] (Fig. 10). The GA structure solution calculation involved 11 variable torsion angles; the peptide groups were constrained to be planar units, with the O–C–N–H torsion angle fixed at 180°. The structure (space group $P4_1$) comprises ribbons that run along the *c*-axis, with adjacent molecules in these ribbons interacting through three N–H···O hydrogen bonds (Fig. 11) in a manner directly analogous to an antiparallel *β*-sheet. Intermolecular N–H···O hydrogen bonds involving the end groups of the oligopeptide chains give rise to two intertwined helical chains running along the 4_1 screw axis.

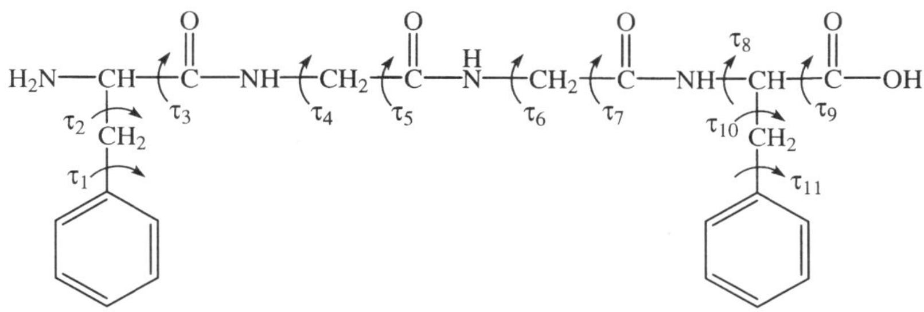

Fig. 10 Molecular structure of Phe–Gly–Gly–Phe, illustrating the torsion angles considered as variables in the GA structure solution calculation

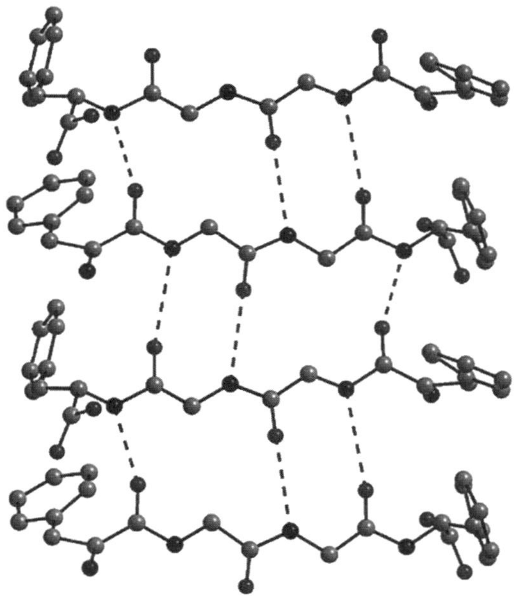

Fig. 11 Final refined crystal structure of Phe–Gly–Gly–Phe, illustrating hydrogen bonding between adjacent molecules

Our next examples concern the characterization of β-turns, which are structural elements that permit polypeptide chain reversals in proteins [65]. Tight turns in proteins and peptides, involving two residues as folding nuclei, have been widely investigated [66–69]. We have applied our GA technique for structure solution of the peptides Piv–^{L}Pro–Gly–NHMe and Piv–^{L}Pro–γ-Abu–NHMe from powder diffraction data, in order to explore the structural properties of these materials (particularly with regard to the formation of β-turns).

In the GA structure solution calculation [70] for Piv–^{L}Pro–Gly–NHMe (Fig. 12), the genetic code comprised nine variables $\{\theta, \phi, \psi, \tau_1, \tau_2, ..., \tau_6\}$ (in space group $P1$, the position $\{x, y, z\}$ of the molecule is fixed arbitrarily). The variable torsion

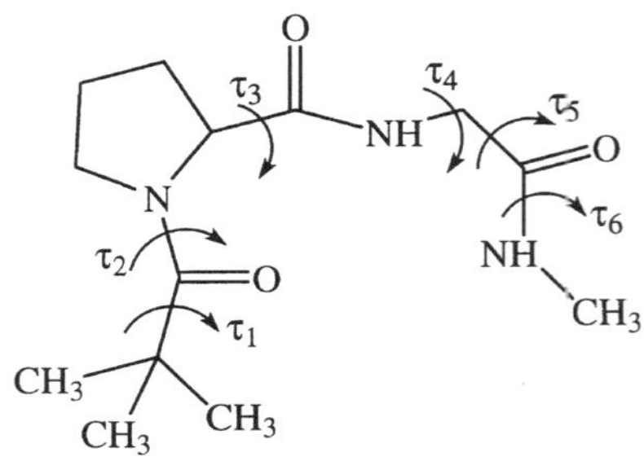

Fig. 12 Molecular structure of Piv–^{L}Pro–Gly–NHMe, illustrating the torsion angles considered as variables in the GA structure solution calculation

angles were allowed to take any value, except τ_2, which was allowed to take only the values 0 or 180°. The O–C–N–H torsion angle between τ_3 and τ_4 was fixed at 180°. The population comprised 100 structures, and in each generation 100 offspring (50 pairs of parents) and 20 mutants were produced.

The final refined structure of Piv–^{L}Pro–Gly–NHMe is shown in Fig. 13, in which it is clear that the molecule adopts a type II β-turn conformation stabilized by an intramolecular 4→1 hydrogen bond between the C=O group of the Piv residue and the methylamide N–H group (N···O 2.99 Å; N···O–C 140.6°). As shown in Fig. 14, adjacent molecules in the crystal structure of Piv–^{L}Pro–Gly–NHMe are linked along the *c*-axis by intermolecular N–H···O hydrogen bonds (N···O 2.87 Å; N···O–C 135.3°).

In view of the fact that Piv–^{L}Pro–Gly–NHMe adopts a classical type II β-turn, it is of interest to explore the influence of introducing additional CH_2 units within the peptide chain, for example, Piv–^{L}Pro–γ-Abu–NHMe (Fig. 15). Again, we have determined the structure of this material [71] directly from powder X-ray diffraction data using the GA technique for structure solution. With one molecule in the asymmetric unit, each structure in the GA calculation was defined by 13 variables (seven variable torsion angles). The torsion angle of the peptide bond of the ^{L}Pro residue was restricted to be either 0° or 180°, and the other

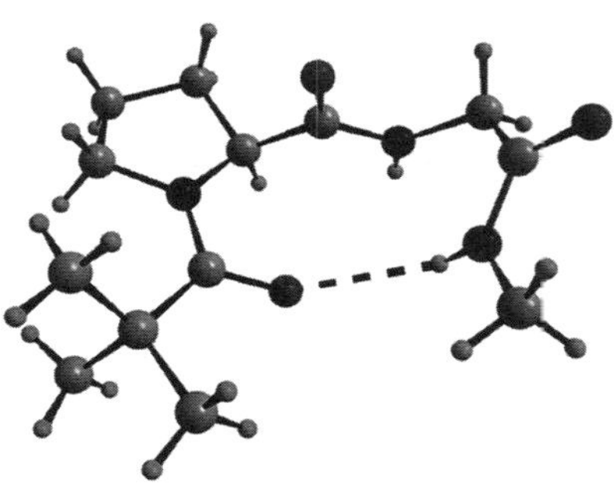

Fig. 13 Conformation of Piv–^{L}Pro–Gly–NHMe in the final refined crystal structure, showing the formation of a type II β turn

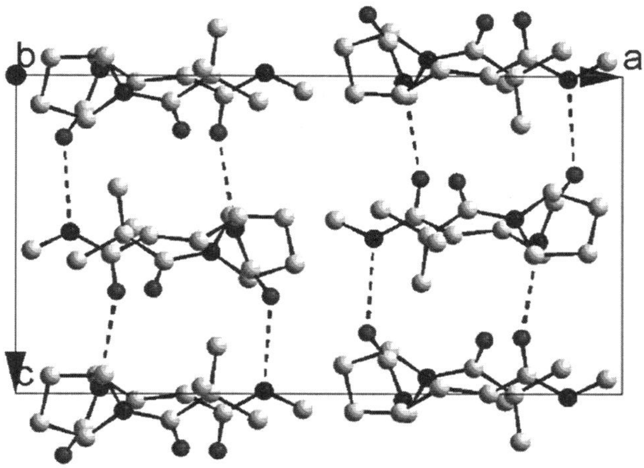

Fig. 14 Illustration of intermolecular N–H···O hydrogen bonds along the *c*-axis in the crystal structure of Piv–^{L}Pro–Gly–NHMe

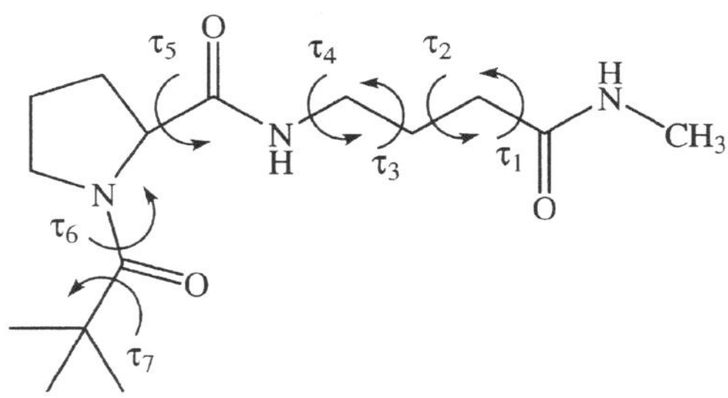

Fig. 15 Molecular structure of Piv–^{L}Pro–γ-Abu–NHMe, illustrating the torsion angles considered as variables in the GA structure solution calculation

two amide linkages CO–NH were maintained as planar units with the O–C–N–H torsion angle fixed at 180°. All other torsion angles were treated as variables.

The GA calculation involved the evolution of a population of 100 structures, and in each generation, 50 mating operations (to produce 100 offspring) and 20 mutation operations were carried out. Figure 16 shows that Piv–^{L}Pro–γ-Abu–NHMe adopts a folded conformation in the crystal structure, reminiscent of chain reversals found in α-peptide structures. In particular, a short C–H···O interaction (H···O 2.51 Å; C···O 3.59 Å; C–H···O 172.4°; hydrogen atom position normalized according to standard geometries from neutron diffraction) is observed between one of the α-methylene hydrogen atoms of γ-Abu and the C=O group of the Piv residue. This C–H···O interaction defines an intramolecular cyclic 10-atom motif, similar to that observed in the classical β-turn [66, 72]. In the cyclic conformation adopted by Piv–^{L}Pro–γ-Abu–NHMe in the crystal structure, all C=O groups point outwards from one face of the molecule and all

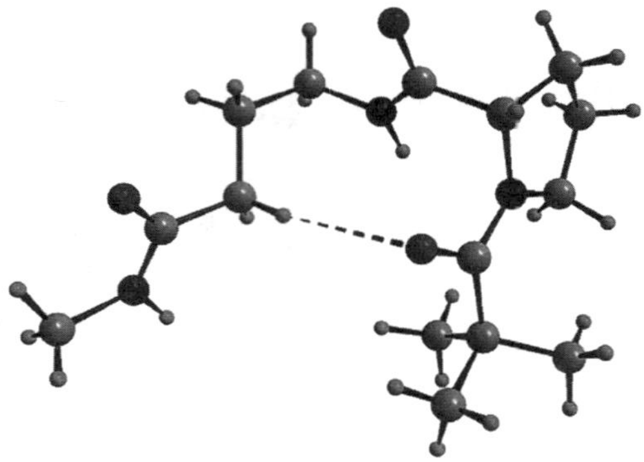

Fig. 16 Conformation of Piv–^{L}Pro–Gly–NHMe in the final refined crystal structure, showing the formation of an intramolecular C–H···O=C hydrogen bond

N–H groups point outwards from the other face, in a manner that allows hydrogen-bonded columns of molecules to be formed along the *c*-axis, with adjacent molecules interacting via two intermolecular N–H···O hydrogen bonds.

3.4 Analysis of the Evolutionary History in Genetic Algorithm Structure Solution Calculations

A general strategy has been developed recently [73] for analysis of the evolutionary events that occur during a GA structure solution calculation, leading to new insights into the evolutionary behaviour in such calculations. Previously, analysis of the performance of a GA structure solution calculation was typically based on consideration of the EPP, which shows the average (R_{ave}), lowest (R_{min}) and highest (R_{max}) values of R_{wp} in the population as a function of generation number in the GA calculation. A typical EPP for structure solution of the α phase of L-glutamic acid using the Lamarckian GA is shown in Fig. 17. The EPP clearly shows systematic decreases in R_{ave} and R_{min} as the calculation progresses, and from such plots it is possible to speculate on several aspects of the performance of the GA calculation. However, the EPP only provides averaged information on the progress of the GA calculation and cannot provide any information on which specific evolutionary events lead to improvements in the quality of the population, nor, most importantly, on which specific evolutionary processes lead towards the correct structure solution. For these reasons, and to allow considerably more insight to be gained into the factors that lead to evolutionary progress during a GA calculation, we have introduced two further means of analysing the evolutionary history in GA calculations – the Evolutionary Distribution Plot (EDP) and the Evolutionary Trajectory Plot (ETP).

The EDP displays the set of R_{wp} values for all members of the population in each generation, and distinguishes the members of the population that arose from mating events, the members of the population that arose from mutation events, and the members of the population that were passed unchanged from the previous

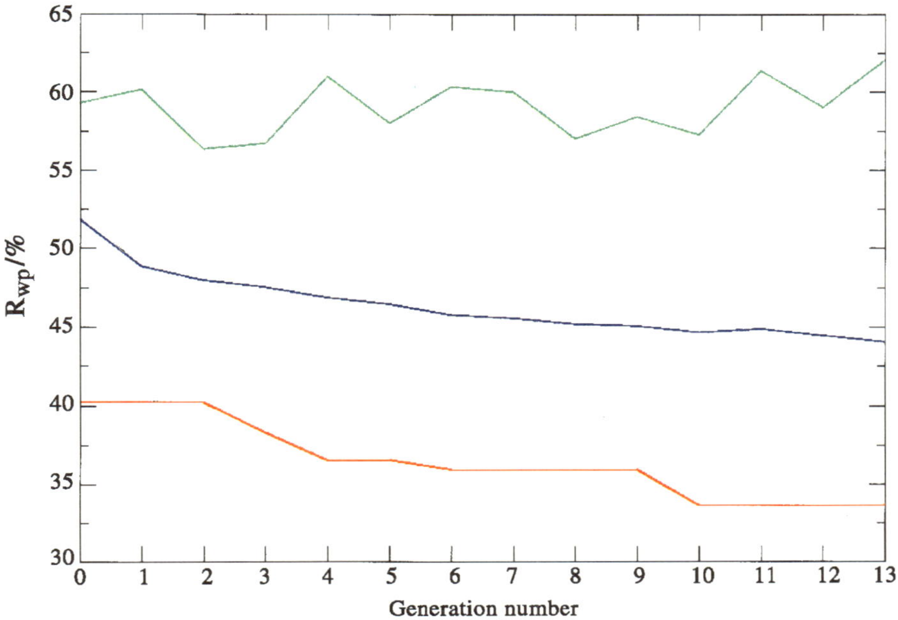

Fig. 17 Evolutionary Progress Plot for the GA structure solution of the α phase of L-glutamic acid. (R_{max}, green; R_{ave}, blue; R_{min}, red)

generation. Clearly the EDP provides substantially more detailed information concerning the evolution of the population than does the EPP (indeed, the EPP may be considered to represent a simplified version of the EDP, representing only the limiting and averaged values of R_{wp}). A typical EDP for structure solution of the α phase of L-glutamic acid is shown in Fig. 18. This EDP shows, inter alia, that a wide distribution of R_{wp} values is represented throughout the evolution, indicating that the population does not become dominated by multiple copies of a particular structure (which would represent stagnation of the population, and is undesirable in an evolutionary context). Nevertheless, it is clear that the population has a distinctly bimodal distribution, with mutant structures usually having considerably higher values of R_{wp} than the other members of the population (for which the R_{wp} values show a general decrease as generation number increases).

The ETP for the same GA calculation on the α phase of L-glutamic acid is shown in Fig. 19. The ETP traces the ancestry of the correct structure solution (or any selected structure in the final population), to provide a graphical representation of the sequence of evolutionary events that led to the production of this structure. This representation is analogous in some respects to a family tree in genealogical contexts. The nodes in the ETP represent the structures in a given population which participate in events that lead subsequently to the production of the correct structure solution. A line between nodes indicates that the structures (in adjacent generations) are connected by an evolutionary event, and the colour of the line indicates the nature of this evolutionary event. Importantly,

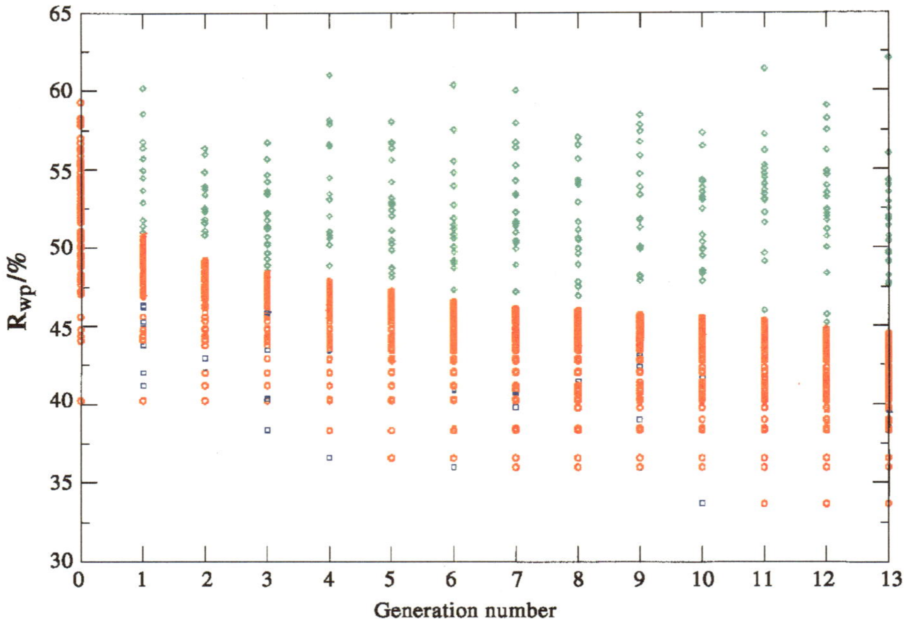

Fig. 18 Evolutionary Distribution Plot for the GA structure solution of the α phase of L-glutamic acid. Structures produced by mating events are shown as blue squares, structures produced by mutations are shown as green diamonds, and structures passing unchanged from the previous generation are shown as red circles

while the EPP and EDP provide information only on the members of the population that are present in each generation, the ETP additionally provides information concerning the types of events that occur in the evolution of the population between successive generations. Analysis of the ETP shown in Fig. 19 provides a detailed understanding of the evolutionary events that are responsible for the production of the correct structure solution, and this type of information may be exploited subsequently in the development of optimized procedures for implementing GA calculations in this field. As an illustration, it is evident from Fig. 19 that a number of the important mating events in the evolution towards the correct structure solution involve mating between a mutant structure and an existing member of the population. The ability of such events to produce improved members of the population arises from the introduction of new genetic material into the mutant structure (which itself may have comparatively high R_{wp}), with this new genetic information then propagated within the population via the mating event, to produce an offspring which typically has low R_{wp}. In recognition of the importance of such coupling between mutation and mating events, we have adopted a strategy of ensuring that all mutant structures survive for at least one generation after their creation, with a high probability (by imposing an artificially high fitness value) that they will be selected as parents in mating events. Several other insights into the crucial steps in the evolutionary pathway towards the correct structure solution may be established from analysis of the ETP, and

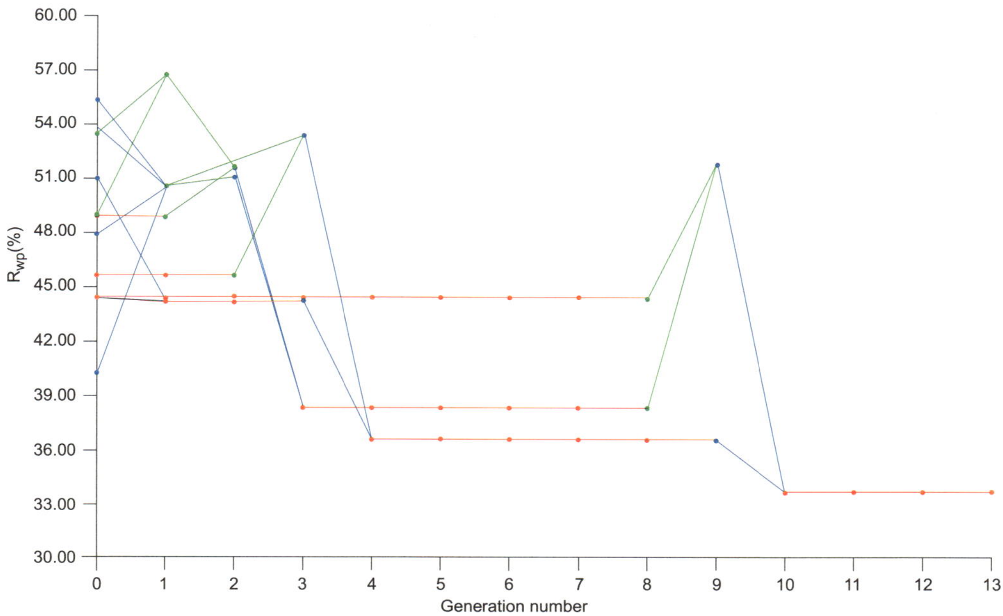

Fig. 19 Evolutionary Trajectory Plot for the GA structure solution of the α phase of L-glutamic acid. In this plot, nodes represent the structures in a given population that participate in evolutionary events that lead subsequently to the production of the correct structure solution. A line between nodes indicates that the structures (in adjacent generations) are connected by an evolutionary event, with the colour of the line indicating the nature of this event. Thus, blue lines represent mating events, green lines represent events involving mutation, and red lines represent structures passing through the natural selection procedure unchanged from the previous generation

may be exploited in the development of improved and optimized implementations of the GA strategy for structure solution.

3.5
Combining Energy Information with Powder Diffraction Data in Structure Solution

In addition to structure determination from powder diffraction data as described earlier, another area of considerable current interest is the computational prediction of crystal structures based on energy simulation techniques. In such work, the potential energy, $E(\mathbf{\Gamma})$, is computed as a function of the set of variables $\mathbf{\Gamma}$ that define the structure (the unit cell and space group are usually also included as variables in such calculations), and the $E(\mathbf{\Gamma})$ hypersurface is searched to find the structure of minimum energy. Representative examples of work in this field may be found in Refs. [74–79], and some work involving the use of evolutionary algorithms to carry out the search procedure has been reported [80–82].

As proposed previously [46], there is considerable scope to exploit the fact that $E(\mathbf{\Gamma})$ hypersurfaces (of the type used in energy simulations) and $R(\mathbf{\Gamma})$ hypersurfaces (of the type used in direct-space techniques for structure solution from powder diffraction data) are based on the same variable-space $\mathbf{\Gamma}$, by defining a new hybrid hypersurface $G(\mathbf{\Gamma})$ which blends $E(\mathbf{\Gamma})$ and $R(\mathbf{\Gamma})$ together in an appropriate functional form. Clearly if the hybrid function $G(\mathbf{\Gamma})$ is defined appropriately, the use of the $G(\mathbf{\Gamma})$ hypersurface in structure solution calculations may have significant advantages over the use of the $R(\mathbf{\Gamma})$ hypersurface. In this regard, we have proposed [83] a specific definition of a new hybrid hypersurface $G(\mathbf{\Gamma})$, which is based on combining desirable characteristics from both the $E(\mathbf{\Gamma})$ and the $R_{wp}(\mathbf{\Gamma})$ hypersurfaces.

Specifically, our hybrid function $G(\mathbf{\Gamma})$ is designed to behave as energy when the value of energy is high and to give increasing importance (ultimately absolute importance) to R-factor as lower values of energy are approached. This behaviour is achieved using a weighting parameter, $w(\mathbf{\Gamma})$, that is an appropriate decreasing function of energy. The hybrid figure-of-merit is defined as

$$G(\mathbf{\Gamma}) = [1 - w(\mathbf{\Gamma})]E^{N}(\mathbf{\Gamma}) + w(\mathbf{\Gamma})R^{N}(\mathbf{\Gamma}) \tag{10}$$

where $w(\mathbf{\Gamma})$ is the weighting function, and $E^{N}(\mathbf{\Gamma})$ and $R^{N}(\mathbf{\Gamma})$ denote normalized energy and normalized R-factor, respectively, with $0 \leq w(\mathbf{\Gamma}) \leq 1$, $0 \leq E^{N}(\mathbf{\Gamma}) \leq 1$ and $0 \leq R^{N}(\mathbf{\Gamma}) \leq 1$. The use of normalized functions allows energy and R-factor to be combined in a straightforward and rational manner. Our normalized energy, $E^{N}(\mathbf{\Gamma})$, is defined as

$$E^{N}(\mathbf{\Gamma}) = \frac{\min[E(\mathbf{\Gamma}), E_{high}] - E_{low}}{E_{high} - E_{low}} \tag{11}$$

where E_{high} and E_{low} represent upper and lower limits of energy, and our normalized R factor, $R^{N}(\mathbf{\Gamma})$, is defined as

$$R^{N}(\mathbf{\Gamma}) = \frac{R_{wp}(\mathbf{\Gamma}) - R_{low}}{R_{high} - R_{low}} \tag{12}$$

where R_{high} and R_{low} represent the expected highest and lowest attainable values of R_{wp} for the system of interest. Methods for assessing the values of E_{high}, E_{low}, R_{high} and R_{low} prior to structure solution calculations have been discussed previously [83].

We use the term "guiding function" to refer to figures-of-merit such as $G(\mathbf{\Gamma})$, in which one property (in the present case energy) is used to guide another property (in the present case R-factor) towards its optimal value. The design of $G(\mathbf{\Gamma})$ is such that $E(\mathbf{\Gamma})$ guides the calculation towards regions of variable-space $\mathbf{\Gamma}$ that correspond to energetically plausible structures, with $R_{wp}(\mathbf{\Gamma})$ becoming progressively more important as the criterion for discriminating the correct structure solution as $E(\mathbf{\Gamma})$ becomes lower. As the primary aim of introducing $E(\mathbf{\Gamma})$ is to avoid implausible structures, rather than to introduce a fine discrimination between different plausible structures [which is achieved using $R_{wp}(\mathbf{\Gamma})$], and as $E(\mathbf{\Gamma})$ makes a significant contribution to $G(\mathbf{\Gamma})$ only when energy is high, it is sufficient for the potential-energy function used to compute $E(\boldsymbol{\Gamma})$ to contain only repulsive terms [83], thus reducing the computation time.

To illustrate the guiding function $G(\boldsymbol{\Gamma})$, we make use of a straightforward visualization method that has been developed [84] for providing a one-dimensional representation of multidimensional hypersurfaces. In this method, a parameter q defines a trajectory through the space defined by the set of variables $\boldsymbol{\Gamma}$, and may be chosen such that the trajectory passes through any specified point and in any direction on the hypersurface. To generate the one-dimensional cross-section through the hypersurface, each variable $\gamma_k (k=1, 2, ..., N)$ in the set $\boldsymbol{\Gamma}$ is varied simultaneously according to

$$\gamma_k(q) = \gamma_{k,0} + q(\gamma_{k,0} - \gamma_{k,\mathrm{lim}}) = \gamma_{k,0} + q\Delta_k \,, \qquad (13)$$

where $\gamma_k(q)$ is the value of γ_k at position q in the trajectory, $\gamma_{k,0}$ is the value of γ_k at the chosen reference point, and $\gamma_{k,\mathrm{lim}}$ is the value of γ_k at the starting point of the trajectory. The value of q is varied from −1 to +1, producing linear variation of all variables in the set $\boldsymbol{\Gamma}$. Each variable γ_k in the set $\boldsymbol{\Gamma}$ starts with the value $\gamma_{k,0}-\Delta_k$ at $q=-1$, passes through the reference point $\gamma_{k,0}$ at $q=0$, and finishes with the value $\gamma_{k,0}+\Delta_k$ at $q=+1$ (where $\Delta_k=\gamma_{k,0}-\gamma_{k,\mathrm{lim}}$). A cross-section through the hypersurface is then constructed by calculating the relevant figure-of-merit at each point along the trajectory defined by $\boldsymbol{\Gamma}=\{\gamma_k(q)\}$.

Figure 20 shows the application of this method to produce cross-sections through the $R_{\mathrm{wp}}(\boldsymbol{\Gamma})$, $E(\boldsymbol{\Gamma})$ and $G(\boldsymbol{\Gamma})$ hypersurfaces for the β phase of L-glutamic acid, the crystal structure of which is known. In this case, $G(\boldsymbol{\Gamma})$ was calculated using the weighting function $w(\boldsymbol{\Gamma})=1-\exp[E^{\mathrm{N}}(\boldsymbol{\Gamma})-1]$. The reference point ($q=0$) is the global minimum on the R_{wp} hypersurface (corresponding to the known crystal structure), and the starting point ($q=-1$) for the cross-section was chosen at random.

Underlying our design of the hybrid figure-of-merit $G(\boldsymbol{\Gamma})$, the main characteristics that we require are

1. That the global minimum in $G(\boldsymbol{\Gamma})$ must correspond to the global minimum in $R^N(\boldsymbol{\Gamma})$.
2. That the basin of attraction around the global minimum in $G(\boldsymbol{\Gamma})$ should be substantially wider than the basin of attraction around the global minimum in $R^{\mathrm{N}}(\boldsymbol{\Gamma})$, and should ideally reflect the broad basins of attraction that are typical on the $E^{\mathrm{N}}(\boldsymbol{\Gamma})$ hypersurface.
3. That the local minima in $G(\boldsymbol{\Gamma})$ should correspond to regions of variable space in which both $E^{\mathrm{N}}(\boldsymbol{\Gamma})$ is low and in which there is a local minimum in $R^{\mathrm{N}}(\boldsymbol{\Gamma})$.
4. That $G(\boldsymbol{\Gamma})$ should be high in regions for which $E^{\mathrm{N}}(\boldsymbol{\Gamma})$ is high.

Characteristic 1 is clearly obligatory if structure solution based on exploring the $G(\boldsymbol{\Gamma})$ hypersurface is to be able to lead to the correct structure solution, and characteristic 2 clearly enhances the probability that the correct structure solution will be found [e.g. in the context of a Lamarckian GA calculation, any structure produced within the broad basin of attraction in $G(\boldsymbol{\Gamma})$ will lead directly, following minimization, to the correct structure solution]. Characteristic 3 ensures that physically plausible structures (i.e. with low energy) that also correspond to local minima in R-factor are given importance in the structure

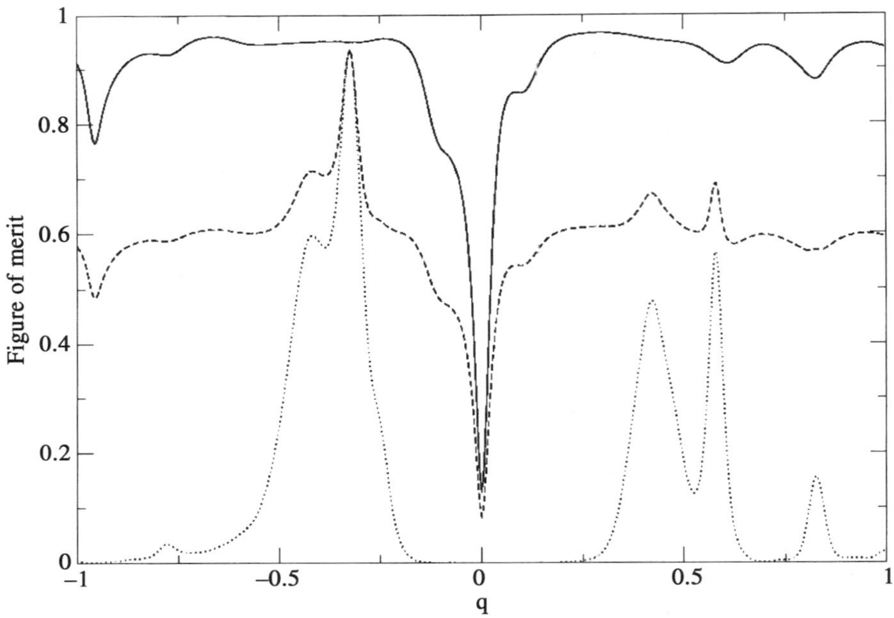

Fig. 20 Illustration of cross-sections through the $R_{wp}(\mathbf{\Gamma})$ (*solid line*), $E(\mathbf{\Gamma})$ (*dotted line*) and $G(\mathbf{\Gamma})$ (*dashed line*) hypersurfaces for the β phase of L-glutamic acid. The same cross-section is shown in each case

solution calculation – while such structures are not necessarily close to the correct structure solution, they may nevertheless contain some good structural characteristics (i.e. good genetic material, in the context of a GA calculation), which may be used productively in the search for the correct structure solution. Finally, characteristic 4 ensures that the direct-space search avoids regions of variable space that correspond to physically implausible (i.e. high-energy) structures. It is clear from inspection of Fig. 20 that the weighting function used to calculate $G(\mathbf{\Gamma})$ satisfies these criteria.

In terms of the application of GA structure solution calculations employing $G(\mathbf{\Gamma})$ as the basis for defining the fitness of members of the population, we find that the progress of the structure solution calculations represents a more systematic and controlled evolution of the population than that typically observed in corresponding calculations using $R_{wp}(\mathbf{\Gamma})$ alone, and this behaviour can be advantageous in avoiding potential problems due to stagnation of the population in GA calculations.

We note that a figure-of-merit with fixed weighting parameter has been employed previously [85] in structure solution from powder diffraction data, although the figure-of-merit used in that paper differs in several respects from $G(\mathbf{\Gamma})$ defined here. The figure-of-merit used in Ref. [85] did not consider normalized energy and normalized R-factor functions, and the R-factor was based on the use of integrated peak intensities rather than a whole-profile fit to the powder diffraction pattern.

3.6 Application of Parallel Computing Concepts in Genetic Algorithms

As already described, the GA approach has proven very successful as a tool for direct-space crystal structure solution from powder diffraction data. However, there remain many opportunities for optimizing GA methodologies. In this section, we discuss one approach for increasing the power and scope of GA methodologies which takes advantage of modern computational techniques.

GAs are implicitly parallel algorithms – the use of a population of trial structures involves simultaneous sampling of several different regions of the hypersurface under investigation, and provides a representative sampling that spans the whole hypersurface. In contrast, in methods based upon weighted random walks across the hypersurface (such as Monte Carlo/simulated annealing approaches), each trial structure is derived from the previous trial structure by small displacements in the structural variables, such that the set of structures sampled forms a Markov chain across the hypersurface. Thus, in such methods, only a very limited region of the hypersurface is actually searched at any given time. Obviously, if the initial trial structure is distant from the global minimum, such a random walk may require a substantial time to locate this minimum, particularly if the dimensionality of the hypersurface is high.

The implicit parallel nature of GAs has been used to great advantage in the development of a related class of algorithms, generically referred to as parallel GAs (PGAs) [86]. Such approaches exploit the parallel nature of GA methods in order to split the overall computational task into several smaller tasks, each of which may be performed on an individual computer processing element. The three main types of PGA may be classified broadly as [86] (1) single-population master-slave PGAs, (2) single-population fine-grained PGAs and (3) multiple-population coarse-grained PGAs. Of these three different approaches to GA parallelization, the multipopulation approach represents the most significant alteration to the standard GA methodology, owing to the introduction of a new evolutionary operation, migration [86]. Thus, the multipopulation approach does not simply divide the existing computational task among different processors, but introduces a new aspect within the evolutionary behaviour.

The GA program EAGER [44–46, 50–52, 59, 60] has recently been parallelized using the multipopulation approach, involving the evolution of N_{sub} different subpopulations, each comprising N_{pop} structures. Within each subpopulation, the operations performed in each evolutionary cycle are identical to those of the single-population GA program EAGER (Fig. 2). However, in addition to these standard evolutionary operations, occasional communications between the N_{sub} subpopulations are also allowed. In general terms, this communication involves migration of trial structures between the subpopulations, resulting in a new set of N_{sub} subpopulations. After the migration process has been completed, the new subpopulations are allowed to continue evolving independently until the next communication operation.

The potential advantages of introducing parallel computational techniques to the standard GA methodology are significant. Most importantly, the introduction of the new migration operation has the potential to inhibit the occurrence of stagnation, which can be a problem in GA calculations. When a population stagnates, a large number of population members become very similar, or even identical, and the resulting reduction of diversity within the population is obviously undesirable as it reduces the ability of the GA to search a wide region of the hypersurface and thus decreases the probability of rapidly locating the global minimum. It is clear that the use of a suitable interaction (i.e. migration) between subpopulations should encourage the maintenance of diversity within each subpopulation by the periodic introduction of new genetic material, and hence should prevent (or at least inhibit) stagnation. This in turn should give rise to greater efficiency in searching the $R_{wp}(\mathbf{\Gamma})$ hypersurface. Furthermore, the improvement in diversity resulting from the migration operation may allow the use of smaller population sizes than those commonly used in the standard GA implementation (recalling that small population sizes tend to be more prone to suffer from stagnation and loss of diversity), with obvious benefits in terms of calculation speed.

In a recent paper [87], we assessed the performance of our multipopulation PGA approach. Two different routines for performing the interpopulation communication process (migration) were tested, as were the effects of subpopulation size. It was found that, for the given test molecule (Piv–^{L}Pro–γ-Abu–NHMe; see Sect. 3.3.3), the performance of the PGA approach was superior to that of the standard serial GA approach with regard to success in locating the global minimum. In particular, it was shown that success rates of 100% could be achieved using the PGA approach, in comparison to an 80% success rate for the same total population size in the standard GA implementation. However, it was also shown that one of the migration routines did not give rise to any significant improvement over the standard GA approach.

There is clearly considerable scope for further optimizing the interpopulation communication routines, and work is currently ongoing in this area. It is expected that improvements in our PGA methodology, as well as optimization of mating and mutation processes, will further extend the scope of the techniques in this field, facilitating the structure solution of molecular materials of greater complexity.

4 Other Applications of Genetic Algorithms in Crystallographic Areas

4.1 Indexing Powder Diffraction Data Using Genetic Algorithms

As discussed in Sect. 3, the first stage of crystal structure determination from powder diffraction data involves determination of the unit cell by indexing the powder diffraction pattern. Clearly it is not possible to proceed with structure solution unless the correct unit cell has been found at this initial stage. Recognizing this issue, a technique employing a GA for indexing powder diffraction data has been reported [88]. The positions of the peaks in a powder diffraction pattern depend on the unit cell dimensions (lattice parameters) $\{a, b, c, \alpha, \beta, \gamma\}$, and the aim of indexing is to determine the correct lattice parameters from knowledge of the peak positions in the experimental powder diffraction pattern. In effect, we require to search a hypersurface $F(a, b, c, \alpha, \beta, \gamma)$ to find the global minimum, where the function F indicates the quality of fit between experimental and calculated peak positions in the powder diffraction pattern. A number of functions may be used for this purpose, including those based on traditional indexing figures-of-merit, such as M_{20} [89]. In our earlier work in this field [88], we instead used a function based on a whole-profile fit to the powder diffraction pattern. We note that another method for indexing powder diffraction data using a GA has also been reported [90].

4.2 Biological Crystallography

In recent years, the field of structural biology has expanded dramatically, with advances in experimental and theoretical crystallography leading to the successful application of X-ray diffraction for the crystal structure analysis of proteins, nucleic acids and other biological macromolecules [91]. It should be noted that there are significant challenges to be faced in the measurement and analysis of the diffraction data from single crystals of such materials, arising in part from the fact that the diffraction data have significant intensity only at low diffraction angles (low resolution) [91]. Of course the sheer size of these macromolecules (typically comprising thousands or tens of thousands of atoms) makes the ab initio structure solution of biomolecules extremely difficult, so most methods for solving the structures involve techniques such as molecular replacement and low-resolution structure solution.

A GA method has been developed [92, 93] for ab initio phasing of low-resolution X-ray diffraction data from highly symmetric structures. The direct-space parameterization used incorporates information on structural symmetry, and has been applied to study the structures of viruses, with resolution as high as 3 Å [93]. A GA has also been introduced [94] to speed up molecular replacement searches by allowing simultaneous searching of the rotational and translational parameters of a test model, while maximizing the correlation coefficient between the observed and calculated diffraction data. An alternative GA for six-dimensional molecular replacement searches has been described [95, 96] and GA methods have also been used [97] to search for heavy atom sites in difference Patterson functions.

A GA has also been developed to determine the low-resolution structures of proteins in solution from small-angle X-ray scattering [98, 99]. Examples of applications include the determination of the low-resolution structures of the active-site cavity of lysozyme, the bilobed structure of γ-crystallin, the shape and size of myoglobin in solution and the horseshoe shape of pancreatic ribonuclease inhibitor [98, 99]. The method involves fitting the small-angle X-ray scattering profile of the protein to a low-resolution bead model, which is optimized using a GA or (in more recent work) an artificial neural network [100]. A similar approach has also been developed by others [101]. These are effectively direct-space or "inverse scattering" approaches, as described in Sect. 3 for solving molecular crystal structures from powder X-ray diffraction data.

4.3
Analysis of Diffuse Scattering from Disordered Materials

The study of disordered materials by analysis of diffuse scattering data is another area that has benefited from the application of evolutionary computing techniques. Work in this area has included the application [102] of a variant of the GA, namely DE [58], to determine disorder parameters from single-crystal X-ray diffraction data. The DE approach differs from standard GA techniques principally in regard to the treatment of crossover and mutation. In DE, a number of mutated structures are first created by adding the differences between corresponding variables of two randomly selected population members to those of a third randomly selected population member. The offspring is then produced by probabilistic selection of variables from another randomly chosen population member and the mutated structure. The offspring structure replaces the parent structure in the population if the offspring structure is judged to be fitter. This single-step crossover/mutation operation based on differences in the variables of population members clearly contrasts with the evolutionary operations utilized in GA methodologies, as described earlier. The deterministic, as opposed to probabilistic, nature of natural selection in DE is a further important difference.

The DE approach was shown [102] to be successful in establishing a good approximation of the disordered crystal structure of the perhydrotriphenylene/1-(4-nitrophenyl)piperazine inclusion compound [102], despite the considerable computational effort required for the simulation of diffuse scattering data. This application is also notable for the use of parallel computing concepts to exploit the implicit parallelism of evolutionary algorithms in order to increase computational efficiency.

4.4
Analysis of Small-Angle X-ray Scattering Data

The DE approach described in Sect. 4.3 has also found application in the study of thin-film structures [103]. In this application, several parameters characterizing thin-film structures, such as layer thickness, layer density and interface depth, were determined by analysis of X-ray reflectivity and small-angle scattering data using a DE algorithm. It was further shown that the technique may be used to infer the presence of surface oxide layers which had been omitted from the thin-film model, as well as allowing estimates of parameters and tolerances of manufactured materials, such as thin-film transistors. A hybrid GA-simulated annealing method has also been used recently [104] to determine the structure of a self-assembled monolayer of decylthiols on the gold (111) surface from grazing-incidence X-ray diffraction data.

4.5 Miscellaneous Applications

In the area of surface science, a GA has been applied [105] for the direct phasing of surface diffraction data as part of a strategy to determine surface structure using direct methods. The method has been applied to both centrosymmetric and noncentrosymmetric two-dimensional structures.

A GA method for determining the Miller indices of crystal faces, using interplanar angles measured experimentally, has been reported [106]. A two-step divide-and-conquer strategy was adopted, and was found to speed up the convergence to the global minimum. The method was tested on crystals of triphenylmethyl chloride, leading to excellent agreement with the results obtained from other indexing methods. A GA has also been used [107] as a search method within crystallographic direct methods for solving surface structure problems.

An evolutionary strategy for refining structural fragments in orientationally disordered crystals has also been introduced [108]. The fragments are embedded into the electron densities which are derived from the experimental data using maximum-entropy techniques [109]. The method has also been used for refinement against structure factors derived from these electron densities. The evolutionary strategy consists only of fitness-based selection and mutation stages (i.e. there is no mating/crossover). Applications include determination of reorientation pathways for the $[PF_6]^-$ anion in KPF_6 and determination of the conformation of the flexible organic molecule 1,3-dioxolane in a sodalite host [108].

Finally, the use of a GA to fit the powder X-ray diffraction patterns of nanocrystalline materials, for example, SiC, GaN and diamond nanocrystals, has also been reported [110].

5 Concluding Remarks

It is clear from the research described in this chapter that evolutionary algorithms have been utilized in optimization problems within a number of crystallographic areas, and there is good reason to forecast that such techniques will find increasing applications in this area of science in the future.

In the present context, it is not appropriate to give a detailed comparison between evolutionary algorithm methods and other methods (such as Monte Carlo and simulated annealing methods) for carrying out global optimization, as the relative merits depend on the particular problem at hand and details of the particular implementation of each approach. Nevertheless, it is important to highlight the fact that evolutionary algorithms operate in an intrinsically parallel manner, in which many different regions of parameter space (corresponding to different members of the population) are investigated simultaneously, and in which the "quality" of different regions of parameter space is directly and actively compared in driving towards populations of continually improving quality. The advantages that arise from the implicit parallel nature of evolutionary algorithms are particularly important when tackling optimization problems defined by a large number of variables. From this viewpoint, and from other considerations, we firmly believe in the advantages offered by evolutionary algorithms for tackling global optimization problems relating to structure determination from diffraction data.

Acknowledgements For collaborations and many useful discussions in connection with our research on structure solution using GAs, we are grateful to Benson Kariuki, Eugene Cheung, Emilio Tedesco, Giles Turner and others mentioned in the references. Our research in this field has been supported by the EPSRC, the University of Birmingham, Purdue Pharma, Ciba Specialty Chemicals, Wyeth, Proctor and Gamble, and Accelrys.

6 References

1. (a) Watson JD, Crick FHC (1953) Nature 171:737, (b) Wilkins MHF, Stokes AR, Wilson HR (1953) Nature 171:738, (c) Franklin RE, Gosling RG (1953) Nature 171:740
2. Perutz MF, Rossman MG, Cullis AF, Muirhead H, Will G, North A (1960) Nature 185:416
3. Dunitz JD (1995) X-ray analysis and the structures of organic molecules. Helvetica Chimica Acta, Basel
4. Glusker JP, Trueblood KN (1985) Crystal structure analysis – a primer. Oxford University Press, Oxford
5. Holland J (1975) Adaptation in natural and artificial systems. University of Michigan Press, Ann Arbor, MI
6. Goldberg DE (1989) Genetic algorithms in search, optimization and machine learning. Addison-Wesley, Reading, MA
7. Cartwright HM (1993) Applications of artificial intelligence in chemistry. Oxford University Press, Oxford.
8. Rietveld HM (1969) J Appl Crystallogr 2:65
9. Young RA (ed) (1993) The Rietveld method. International Union of Crystallography and Oxford University Press, Oxford
10. Christensen AN, Lehmann MS, Nielsen M (1985) Aust J Phys 38:497
11. Cheetham AK, Wilkinson AP (1991) J Phys Chem Solids 52:1199
12. McCusker LB (1991) Acta Crystallogr Sect A 47:297
13. Cheetham AK, Wilkinson AP (1992) Angew Chem Int Ed Engl 31:1557
14. Rudolf PR (1993) Mater Chem Phys 35:267
15. Langford JI, Louër D (1996) Rep Prog Phys 59:131
16. Poojary DM, Clearfield A (1997) Acc Chem Res 30:414
17. David WIF (1987) J Appl Crystallogr 20:316
18. David WIF (1990) Nature 346:731
19. Jansen J, Peschar R, Schenk H (1992) J Appl Crystallogr 25:237
20. Estermann MA, McCusker LB, Baerlocher C (1992) J Appl Crystallogr 25:539
21. Estermann MA, Gramlich V (1993) J Appl Crystallogr 26:396
22. Gilmore CJ, Shankland K, Bricogne G (1993) Proc R Soc London Ser A 442:97
23. Sivia DS, David WIF (1994) Acta Crystallogr Sect A 50:703
24. Gilmore CJ (1996) Acta Crystallogr Sect A 52:561
25. David WIF (1999) J Appl Crystallogr 32:654
26. Harris KDM, Tremayne M, Lightfoot P, Bruce PG (1994) J Am Chem Soc 116:3543
27. Harris KDM, Tremayne M (1996) Chem Mater 8:2554
28. Harris KDM, Tremayne M, Kariuki BM (2001) Angew Chem Int Ed Engl 40:1626
29. Kariuki BM, Zin DMS, Tremayne M, Harris KDM (1996) Chem Mater 8:565
30. Tremayne M, Kariuki BM, Harris KDM (1996) J Appl Crystallogr 29:211
31. Tremayne M, Kariuki BM, Harris KDM (1996) J Mater Chem 6:1601
32. Tremayne M, Kariuki BM, Harris KDM (1997) Angew Chem Int Ed Engl 36:770

33. Elizabé L, Kariuki BM, Harris KDM, Tremayne M, Epple M, Thomas JM (1997) J Phys Chem B101:8827
34. Tremayne M, Kariuki BM, Harris KDM, Shankland K, Knight KS (1997) J Appl Crystallogr 30:968
35. Harris KDM, Kariuki BM, Tremayne M (1998) Mater Sci Forum 278–291:32
36. Newsam JM, Deem MW, Freeman CM (1992) Accuracy in powder diffraction II: NIST special publication no 846, p 80
37. Ramprasad D, Pez GB, Toby BH, Markley TJ, Pearlstein RM (1995) J Am Chem Soc 117: 10694
38. Andreev YG, Lightfoot P, Bruce PG (1996) Chem Commun 2169
39. Andreev YG, MacGlashan GS, Bruce PG (1997) Phys Rev B 55:12011
40. Freeman CM, Gorman AM, Newsam JM (1997) Simulated annealing and atructure solution. In: Catlow CRA (ed) Computer modelling in inorganic crystallography. Academic, San Diego
41. David WIF, Shankland K, Shankland N (1998) Chem Commun 931
42. Engel GE, Wilke S, König O, Harris KDM, Leusen FJJ (1999) J Appl Crystallogr 32:1169
43. Bell AMT, Smith JNB, Attfield JP, Rawson JM, Shankland K, David WIF (1999) New J Chem 23:565
44. Kariuki BM, Serrano-González H, Johnston RL, Harris KDM (1997) Chem Phys Lett 280:189
45. Harris KDM, Johnston RL, Kariuki BM, Tremayne M (1998) J Chem Res 390
46. Harris KDM, Johnston RL, Kariuki BM (1998) Acta Crystallogr Sect A 54:632
47. Shankland K, David WIF, Csoka T (1997) Z Kristallogr 212:550
48. Shankland K, David WIF, Csoka T, McBride L (1998) Int J Pharm 165:117
49. Csoka T, David WIF, Shankland K (1998) Mater Sci Forum 278–281:294
50. Kariuki BM, Calcagno P, Harris KDM, Philp D, Johnston RL (1999) Angew Chem Int Ed Engl 38:831
51. Kariuki BM, Psallidas K, Harris KDM, Johnston RL, Lancaster RW, Staniforth SE, Cooper SM (1999) Chem Commun 1677
52. Harris KDM, Johnston RL, Kariuki BM (1998) An Quim Int Ed 94:410
53. Reck G, Kretschmer R-G, Kutschabsky L, Pritzkow W (1988) Acta Crystallogr Sect A 44:417
54. Cirujeda J, Ochando LE, Amigó JM, Rovira C, Rius J, Veciana J (1995) Angew Chem Int Ed Engl 34:55
55. Dinnebier RE, Stephens PW, Carter JK, Lommen AN, Heiney PA, McGhie AR, Brard L, Smith AB III (1995) J Appl Crystallogr 28:327
56. Hammond RB, Roberts KJ, Docherty R, Edmondson M (1997) J Phys Chem B101:6532
57. Seaton CC, Tremayne M (2002), Chem Commun 880
58. Storn R, Price KV (1997) J Global Optimization 11:341
59. Harris KDM, Johnston RL, Albesa Jové D, Chao MH, Cheung EY, Habershon S, Kariuki BM, Lanning OJ, Tedesco E, Turner GW (2001) EAGER. University of Birmingham [an extended version of the program GAPSS, Harris KDM, Johnston RL, Kariuki BM (1997) University of Birmingham]
60. Turner GW, Tedesco E, Harris KDM, Johnston RL, Kariuki BM (2000) Chem Phys Lett 321:183
61. Lehman MS, Koetzle TF, Hamilton WC (1972) J Cryst Mol Struct 2:225
62. Lehman MS, Nunes AC (1980) Acta Crystallogr Sect B 36:1621
63. Visser JW (1969) J Appl Crystallogr 2:89
64. Tedesco E, Turner GW, Harris KDM, Johnston RL, Kariuki BM (2000) Angew Chem Int Ed Engl 39:4488
65. Richardson JS (1981) Adv Protein Chem 34:167
66. Rose GD, Gierasch LM, Smith JA (1985) Adv Protein Chem 37:1
67. Wilmot CM, Thornton JM (1988) J Mol Biol 203:221
68. Toniolo C (1980) CRC Crit Rev Biochem 9:1
69. Gunasekaran K, Gomathi L, Ramakrishnan C, Chandrasekhar J, Balaram P (1998) J Mol Biol 284:1505

70. Tedesco E, Harris KDM, Johnston RL, Turner GW, Raja KMP, Balaram P (2001) Chem Commun 1460
71. Cheung EY, McCabe EE, Harris KDM, Johnston RL, Tedesco E, Raja KMP, Balaram P (2002) Angew Chem Int Ed Engl 41:494
72. Venkatachalam CM (1968) Biopolymers 6:1425
73. Habershon S, Turner GW, Harris KDM, Johnston RL, Johnston JM (2002) Chem Phys Lett 353:185.
74. Karfunkel HR, Gdanitz RJ (1992) J Comput Chem 13:1171
75. Price SL (1997) In: Catlow CRA (ed) Computer modelling in inorganic crystallography. Academic, San Diego, p 269
76. Gdanitz RJ (1997) In: Gavezotti A (ed) Theoretical aspects and computer modelling of the molecular solid state. Wiley, New York, p 185
77. Gavezotti A (1998) Crystallogr Rev 7:5
78. Verwer P, Leusen FJJ (1998) Rev Comput Chem 12:327
79. Lommerse JPM, Motherwell WDS, Ammon HL, Dunitz JD, Gavezotti A, Hoffman DWM, Leusen FJJ, Mooij WTM, Price SL, Schweizer B, Schmidt MU, Van Eijck BP, Verwer P, Williams DE (2000) Acta Crystallogr Sect B 56:697
80. Bush TS, Catlow CRA, Battle PD (1995) J Mater Chem 5:1269
81. Woodley SM, Battle PD, Gale JD, Catlow CRA (1999) Phys Chem Chem Phys 1:2535
82. Woodley SM (2004) Prediction of Crystal Structures Using Evolutionary Algorithms and Related Techniques, in this book
83. Lanning OJ, Habershon S, Harris KDM, Johnston RLJ, Kariuki BM, Tedesco E, Turner GW (2000) Chem Phys Lett 317:296
84. Turner GW, Tedesco E, Harris KDM, Johnston RL, Kariuki BM (2001) Z Kristallogr 216:187
85. Putz H, Schön JC, Jansen M (1999) J Appl Crystallogr 32:864
86. Cantú-Paz E (1998) A survey of parallel genetic algorithms. In: Calculateurs paralleles, reseaux et systems repartis, vol 10. Hermes, Paris, pp 141–171
87. Habershon S, Harris KDM, Johnston RL (2003) J Comput Chem 24:1766
88. Kariuki BM, Belmonte SA, McMahon MI, Johnston RL, Harris KDM, Nelmes RJ (1999) J Synchrotron Radiat 6:87
89. De Wolff PM (1968) J Appl Crystallogr 1:108
90. Paszkowicz W (1996) Mater Sci Forum 228–231:19
91. Glusker JP, Lewis M, Rossi M (1994) Crystal structure analysis for chemists and biologists. Wiley-VCH, New York
92. Miller ST, Hogle JM, Filman DJ (1996) Acta Crystallogr Sect D 52:235
93. Miller ST, Hogle JM, Filman DJ (2001) J Mol Biol 307:499
94. Chang G, Lewis M (1997) Acta Crystallogr Sect D 53:279
95. Kissinger CR, Gehlhaar DK, Fogel DB (1999) Acta Crystallogr Sect D 55:484
96. Kissinger CR, Gehlhaar DK, Smith BA, Bouzida D (2001) Acta Crystallogr 16:1460
97. Chang G, Lewis M (1994) Acta Crystallogr Sect D 50:667
98. Chacon P, Moran F, Diaz JF, Pantos E, Andreu JM (1998) Biophys J 74:2760
99. Chacon P, Diaz JF, Moran F, Andreu JM (2000) J Mol Biol 299:1309
100. Wriggers W, Chacón P (2001) J Appl Crystallogr 34:773
101. Zipper P, Durchschlag H (2002) Physica A 314:613
102. Weber T, Bürgi H-B (2002) Acta Crystallogr Sect A 58:526
103. Wormington M, Panaccione C, Matney KM, Bowen DK (1999) Philos Trans R Soc London Ser A 357:2827
104. Vargas MC, Torres-Jimenez J, Ruiz-Suarez JC (2002) J Phys A 35:3865
105. Landree E, Collazo-Davila C, Marks LD (1997) Acta Crystallogr Sect B 53:916
106. Tam KY, Compton RG (1995) J Appl Crystallogr 28:640
107. Marks LD, Erdman N, Subramanian A (2001) J Phys Condens Matter 13:10677
108. Knorr K, Madler F (1999) J Appl Crystallogr 32:902
109. Knorr K, Madler F, Papoular RJ (1998) Microporous Mesoporous Mater 21:353
110. Pielaszek R, Gierlotka S, Stelmakh S, Grzanka E, Palosz B (2002) Defect Diffus Forum 208:187

Structure and Bonding, Vol. 110 (2004): 95–132
DOI 10.1007/b13934

Prediction of Crystal Structures Using Evolutionary Algorithms and Related Techniques

Scott M. Woodley

Davy Faraday Research Laboratory, The Royal Institution of Great Britain, 21 Albemarle Street, London W1S 4BS, UK
E-mail: smw@ri.ac.uk

To my wife Jacquelyn Sarah and our children Samson Blake, Maximus Aidan and Jedidiah Benedict

Abstract Methods, evolutionary and systematic search approaches, and applications of crystal structure prediction of closest-packed and framework materials are reviewed. Strategies include developing better cost functions, used to assess the quality of the candidate structures that are generated, and ways to reduce the set of candidate structures to be assessed. The crystallographic coordinates for new materials, available only as a powder sample, are often intractable from diffraction data alone. In recent years, steady progress has been made in the ability to solve previously unknown crystal structures of such compounds, the generation of known structures (inferring more confidence in such approaches) and the prediction of hypothetical yet-to-be-synthesised structures.

Keywords Genetic algorithms · Simulated annealing · Structure prediction · Cost function · Exclusion zones

1 **Introduction** 96

2 **Structure Prediction of Inorganic Materials** 97

2.1 Generating Candidate Structures 97
2.1.1 Introduction 97
2.1.2 Random Start – Move Class Operators 99
2.1.3 Systematic Search 104
2.2 Evaluating Candidate Structures – Cost Function 106
2.2.1 Introduction 106
2.2.2 Cost-Function Definitions 107
2.2.3 Switching Cost Functions 109

3 **Applications of Crystal Structure Prediction** 111

3.1 Genetic Algorithms – Closest-Packed Oxides 111
3.1.1 Multistage Scheme, Examples 111
3.1.2 Polymorphism 115
3.2 Simulated Annealing – Closest-Packed Structures 117
3.2.1 Fixed Unit Cell Parameters 117
3.2.2 Unit Cell and Coordinates Unknown 121

3.3 Framework Crystal Structures . . . 124
3.3.1 Genetic Algorithm Approach . . . 124
3.3.2 Simulated Annealing Approach . . . 126

4 **Conclusions and Perspectives** . . . 128

5 **References** . . . 130

Abbreviations

GA	Genetic algorithm
SA	Simulated annealing
EZ	Exclusion zone
N_i	Number of atoms/ions per unit cell
N_G	Number of grid points across unit cell
R_n	Random number between 0 and 1
T	Temperature parameter
P_x	Probability for event x
R	Discrepancy between simulated and observed diffractograms
r_{ij}	Distance between atoms/ions i and j
q_i	Charge on ion i
Cn_i	Coordination number for ion i
A, B, C, ρ, R	Interatomic potential parameters (for interaction between ions i and j)
N, M, L, m, i	Integer numbers
k, λ	Real numbers

1 Introduction

Computer simulation techniques have acquired a truly predictive capability in structural chemistry. Significant progress in developing then applying both the theory and the computer software has been made in two different aspects: (1) to assist in the structural solution where it is not always possible to grow a large enough crystal to determine the crystallographic coordinates of a new compound from its X-ray diffraction pattern alone [1, 2]; (2) to predict new feasible compounds that have not yet been synthesised [3–5] or that do not exist under physical conditions obtainable in the laboratory [6]. Once the crystal structure is known one can then investigate and predict the properties of the compound [7–10]. For example, a computational static lattice package, like GULP [11, 12], can be used to calculate the activation energy for the migration of defects, the vibrational frequencies, etc., or a more computationally expensive electronic structure package, like CASTEP [13], can be utilised to obtain detailed information on chemical bonding.

Historically, the application of computer modelling focused on reproducing known structures by employing interatomic potentials and minimising the lattice energy [14]. Next the focus of computational inorganic chemists turned to refining approximate models and to the application of ab initio methods to investigate electronic-dependent properties. The work of Shannon et al. [15] provides an excellent illustration of how a postulated trial structure that has been improved by lattice-energy minimisation can serve as the starting point for the refinement of a structure using high-resolution synchrotron X-ray powder diffraction patterns. With the rise of ab initio methods, when interatomic potential parameters are refined observed structural data have been supplemented with simulated data [16]. Note that even when electronic methods are required, it is often beneficial to start a preliminary study which employs interatomic potentials as this technique is always far less computationally demanding, both in computing time and memory requirements. Nowadays, ambitiously, computational methods have been applied to the prediction of structures with no initial knowledge of the crystallographic coordinates and, if necessary, the unit cell dimensions. This difficult problem [17] is essentially one of global optimisation; the challenge is to develop procedures that can generate plausible approximate models for a crystal structure, which may be subsequently refined by energy-minimisation methods. When there are a large number of unknowns, and therefore a vast number of possible trial structures that could be generated, the science of crystal structure prediction will depend on how to best evaluate the trial structures and how to avoid evaluating too many of the unwanted structures. As we will see, when employing a genetic algorithm (GA), typically over 90% of the computing time is spent evaluating the quality of trial structures, or "candidates" [18], where the relationship between quality and energy hopefully implies that the lower the lattice energy for a trial structure the better the candidate. Even with today's increase in computer power, the simplest way to calculate the lattice energy may prove too expensive.

In the next section the ideas behind several methods, including the GA and a simulated annealing (SA) approach [19, 20], then their implementation used to generate ionic crystal structures are reviewed. This will contain an introduction to the types of "move class operators" and the various types of "cost functions" used to modify the current trial structure(s) and to assess the quality of the trial structures, respectively. In the third section recent applications of the GA and SA approaches to closest-packed ionic systems and then to open-framework crystal structures are reviewed.

2 Structure Prediction of Inorganic Materials

2.1 Generating Candidate Structures

2.1.1 *Introduction*

The structure of a new crystalline compound is determined, or solved, once the dimensions of the unit cell, its contents and their relative positions are known. Typically, new compounds are synthesised in the form of a powder (polycrystalline) and as a consequence the crystal structure is not readily obtainable. From

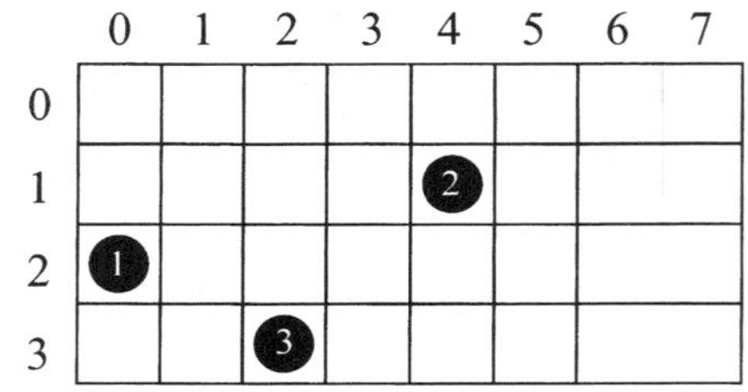

Fig. 1 2D example of a uniform grid applied to the unit cell. The ions are on grid points 02 41 23

the X-ray powder diffraction pattern, chemical analyses and knowledge of the synthetic method the dimensions of the unit cell and the contents may be obtained directly but not the atomic/ionic relative positions (coordinates). However, from trial structural solutions (candidate structures) it is relatively straightforward to generate the simulated X-ray powder diffraction patterns. If there is a suitable match between the simulated and observed diffractograms then the particular candidate structure corresponds to the desired solution or the correct structure of the compound. In practice only approximate coordinates are required as the candidate structure can be refined using a Rietveld method [21–24] to improve the fit between the simulated and observed diffractograms. The task in hand is to predict the approximate coordinates of the constituent ions within the unit cell.

But how difficult is it to search through all configurations with the restrictions of a fixed unit cell and a number of constituent ions, N_i? One approach is to discretise the unit cell such that the ions can only occupy one of N_G grid points (Fig. 1). If all N_i ions are chemically different and ions are not forced to sit on unique grid points then there are $N_G^{N_i-1}$ possible candidate structures. Choosing the spacing of the grid points to be defined by the accuracy required, 0.2 Å say, and assuming a cubic unit cell with an edge of length of 4.0 Å, say, then we would have 8,000 grid points per unit cell. Using these parameters to predict a closest-packed arrangement where N_i=8 and assuming that our computer can evaluate 10,000 candidate structures per second, then we would need over 10^{15} years to evaluate all candidate structures. However, if four of the constituents are sodium ions and the remainder are chloride ions, then a GA can be employed to predict the rock salt structure within minutes [1]. Obviously, due to having chemically identical ions within the unit cell, the problem has been reduced as more than one of the $N_G^{N_i-1}$ possible candidate structures gives the correct structure. But importantly the correct candidate structure was found without evaluating all possible candidate structures.

There are two quite different approaches to crystal structure prediction, but both methods are attempts to reduce the number of candidate structures to evaluate before the desired structure is found. The first modifies a trial structure or set of structures using either a SA or a GA approach, where random ionic coordinates (unknowns, which need not include all ions within the unit cell and may also include the unit cell dimensions) are used to create the initial candidate structure(s). The second is to systematically evaluate a subset of candidate structures that obey certain rules of connectivity and highlight those that may be realised in nature [25–27]. One common thread with all these methods is that the quality of the candidate structures generated is assessed by evaluating a cost function (lattice or free energy, say). The main principle used, here and in the applications discussed later, is that stable (or metastable) structures are those with a lattice energy (cost function) that is at a local minimum on the lattice-energy hypersurface. Thus, the candidate structures generated are usually refined by directly minimising the lattice energy.

In the SA approach, the ideas of annealing are used. When a metal melts, the atoms are in a disordered state. By annealing, cooling slowly from the molten state, the atoms have a chance to crystallise in an ordered manner, with the lattice energy reaching the global minimum. If the temperature is decreased too rapidly (the system is quenched) or if the initial temperature is too low, then the crystalline system will contain (unwanted) defects such as vacancies and/or dislocations. That is, the lattice energy of the system becomes trapped in a local minimum. This process is mimicked in the simulations where the ions hopefully settle to produce the desired structure (corresponding to the global minimum of the lattice energy) as an appropriately defined temperature parameter is reduced. Note that the configuration of ions (candidate structure) at each temperature is continually perturbed according to certain prescribed rules (move class operators); see later.

The GA approach is based on ideas taken from Darwin's theory of evolution. A "population" of candidate structures is created. Then, competition to "procreate" in the population is simulated. That is, information from the better candidate structures of the current population is statistically more likely to be chosen to create (using a suitable move class) candidate structures for a new population. By repeating this process, candidate structures containing good/bad features will prosper/diminish and a candidate structure should evolve which best fits the environment (fixed constraints of the system including the definition of the cost function). Here, the phrase "best fits" refers to the lattice energy corresponding to the global minimum. The move class operators of the SA and GA approaches are now described.

2.1.2
Random Start – Move Class Operators

From candidate structures where the ionic coordinates are initially randomised, both SA and the GA require move class operators which can generate new candidate structures from the current candidate structure(s). The success of both methods is dependent upon the definitions of these operators. It is important that the desired structures are accessible using the move class operators.

When applying a SA approach to crystal structure prediction, a Metropolis Monte Carlo scheme [20], rather than molecular dynamics [28], is usually chosen to sample the configurational space (different possible candidate structures). In practice, this scheme proceeds by comparing the quality (value of the cost function) of a new candidate structure with the current candidate structure. The new candidate is either rejected or used to replace the current candidate structure depending on whether a random number, R_n, between 0 and 1, is greater than or less than $e^{-\Delta E/T}$. Here, ΔE is the difference in the cost-function value between the new and current candidate structures and T is the temperature parameter used to control the likelihood of crossing a barrier between different regions of the cost function containing different local minima. By repeating this process and gradually reducing T, a random walk across the cost-function landscape is taken, where the probability of accepting only better quality candidate structures increases. Eventually the candidate structure is quenched to the corresponding local cost-function minimum. The main move class operator "move a randomly chosen ion", which is used to create the next candidate structure, is applied one or more times to the current candidate. Other move class operators include "swap two randomly chosen ions", "remove/add ions" and "change unit cell dimensions". Allowing the unit cell dimensions to change will be discussed later. For simplicity, in the rest of this section the unit cell dimensions and contents are assumed to be known and thus fixed.

The GA has two important move class operators, "crossover" and "mutate". To explain these operators we recall that all the information required to build any organism is encoded within its DNA. On "reproduction", DNA from both parents combines in a crossover process with, occasionally, a few mutations to produce the DNA of the offspring. As a result of direct competition, or evolution, DNA representing the most suitable organism for its "environment" is eventually produced (the Darwinian principle). In our analogous system the fixed parameters (cost function and, for example, unit cell with space group $P1$) form the environment and the concatenated sequence of unknown variables (crystallographic coordinates) is our DNA. By fixing the order in which the variables are concatenated (and assuming all ions are chemically different) there exists a one-to-one relationship between each possible candidate structure and the simulated DNA. Note that the way in which the candidate structures are assessed forms part of the environment as any change in the definition of the cost function may influence the outcome (predicted structure) of the GA.

Although there are different definitions of the GA move class operator mutate, the purpose of mutate is the same; to prevent the population of candidate structures becoming a population of similar candidate structures (to maintain the "diversity" of the population). Likewise there are also different definitions for the GA move class crossover. In one definition mutate is the process of randomly displacing one ion within a candidate structure and crossover is the process of swapping a random number of ionic coordinates in the simulated DNA of two candidate structures within the current population. Note that in the crossover process the nth variable of one simulated DNA sequence is swapped with the nth variable of the other.

In another popular definition of mutate and crossover, a binary representation of the unknown variables is required whereby the simulated DNA is converted into a concatenated sequence of binary numbers (0's and 1's). To obtain a binary representation of the ionic coordinates, the ions are constrained in that they can only sit on one of 2^m discrete grid points across the unit cell (Fig. 1). For each grid point there is a unique binary number of length m. Note that the grid points can either be numbered 0 to 2^m-1 (000 to 111 for $m=3$) or, as shown in Fig. 1, have each of the components of the grid coordinates numbered from 0 to $2^{m'}-1$ (where m' can be different for each component). Combined with an evenly spaced grid, this latter method of numbering the grid points is particularly convenient when the fractional coordinates of the ions are required. By concatenating the binary representation of the components (each of length m') a unique sequence of length m per grid point is obtained, although when comparing the two methods the series of 0's and 1's for a particular grid point may differ. The binary representation of the simulated DNA is created by concatenating the binary representation of the unknown ionic coordinates. Again, by fixing the order of the N_i constituent ions (and assuming all ions are chemically different), the simulated DNA for each different candidate structure will have a unique sequence, of length mN_i, of 0's and 1's. For example, the DNA sequence for the example given in Fig. 1 (where a 2D unit cell is used for clarity) would be 000 10 100 01 010 11 when the grid coordinates are numbered from (0,0) to (7,3). In the discussions that follow, this type of binary representation, as opposed to numbering all the grid points from 0 to N_G-1, is assumed.

Mutate now involves randomly changing a few bits (0's to 1's or 1's to 0's) along the binary representation of the simulated DNA sequence. This has the effect of displacing one or more atoms within the unit cell (Fig. 2). In practice a random number R_n, between 0 and 1, is generated for each bit in the simulated DNA sequence and is compared to the GA parameter P_m. P_m is generally set such that on average one bit is changed per candidate structure. Thus, mutate will typically create a new candidate that has an ion which has moved 1, 2, 4, 8, 16,…,$2^{m'}$ grid points. That is, the fractional ionic coordinate changes by $^1/_{m'}, \ldots, ^1/_4, ^1/_2$. This selection of possible displacements may be desirable and has few, although still possible, larger jumps across the unit cell. One concern with the definition of the chosen set of move class operators is that the desired crystal structure can be created (and as easily as possible). The definition of one mutation using this binary representation does not allow, for example, an ion to be displaced by 13 grid points. Using three mutations, it is possible in two different ways. Although on average one mutation occurs per candidate structure, three mutations per candidate structure could occur. The mutations occur over several generations. The latter requires that any intermediate candidate structure can prosper (survive for procreation). If this intermediate candidate structure is better/worse than the original then it will be easier/harder to find the desired crystal structure by relying on three mutations per candidate structure to occur. The definition of the

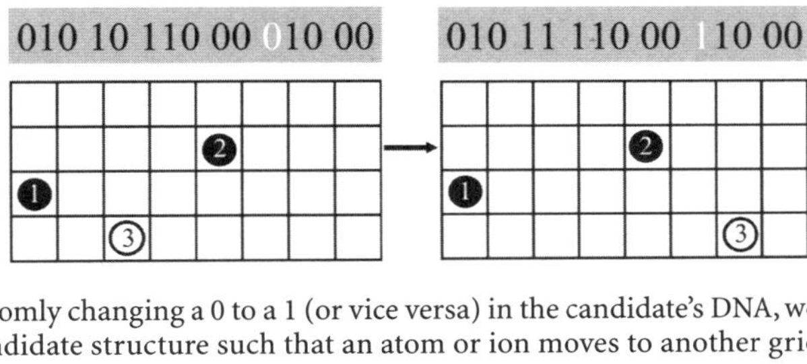

Fig. 2 By randomly changing a 0 to a 1 (or vice versa) in the candidate's DNA, we can effectively mutate the candidate structure such that an atom or ion moves to another grid point

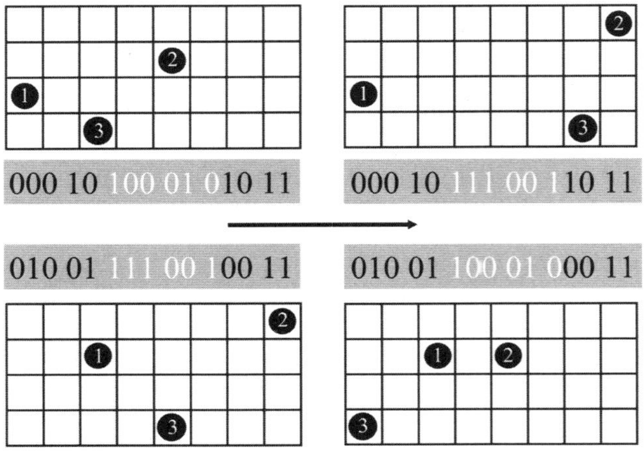

Fig. 3 By swapping a section (random length and position within DNA) of 1's and 0's between two DNA strands, we can generate two new DNA strands. Each DNA strand represents a candidate structure

cost function (which forms part of the environment) will affect the outcome and will be discussed later.

Whereas, at a given stage in the calculation, SA only has one current candidate structure, a GA has a current population of candidate structures; thus, after selecting two candidates the process of crossover can be applied to generate two new candidate structures. For example, in Fig. 3, two new candidate structures have been created by applying a "two-point" crossover; randomly choosing two points along the sequence of 0's and 1's of the simulated DNA and swapping the middle section of one candidate structure with an equivalent section of another candidate structure. Note that crossover in the binary representation can create different ionic coordinates than those contained within the original two candidate structures (ion 3 in Fig. 3). This is due to the fact that the random points where the sequence is cut need not be between the binary numbers which were concatenated. The effect this crossover procedure has on the diversity of the population depends on how the diversity is defined. For example, if the diversity is based on the number of different coordinates for each ion in the population then the diversity may increase, whereas if the diversity is defined as the number of bits along the simulated DNA where 0 and 1 both exist in the current population (as used later) then the diversity does not change.

Of course there is no reason for swapping just one section. Using the binary representation of the simulated DNA, we can vary the number and length of the random sections of 0's and 1's. Whether swapping one or many sections is more beneficial will depend on the environment, or indeed the stage (earlier or later populations) of the GA [29]. Some experimenting with the GA parameters/move class operators is always advisable.

In practise, for each pair of candidates that is chosen for procreation, crossover is applied if R_n is less than the GA parameter P_c (generally set between 0.4 and

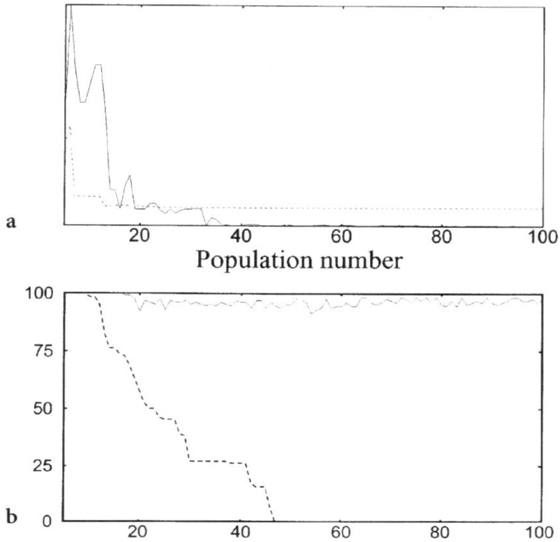

Fig. 4 A typical change in **a** the cost function (arbitrary units) for the best candidate out of a population of 100 candidates and **b** the diversity (percent) of the population where a genetic algorithm (*GA*) is used to predict the structure of BaO with N_G=64^3, N_i=8, P_t=0.9, P_c=0.4 and P_m=0.0 (*broken line*) or $^1/_{126}$ (*solid line*)

0.5). By implementing the crossover and mutate move class operators, we can produce "children" candidate structures containing slightly modified strands of DNA from two "parent" candidate crystal structures. However, for example, if P_c and P_m are both close to zero then one "child" per pair will probably look like its "mother" and the other like its "father". Occasionally one of the new candidate structures will contain the better features contained in the two original candidate structures and the other the worse features. Note that the selection process of choosing which of the current population of candidate structures are paired up to generate the new population of candidate structures is weighted by their respective quality, or "fitness" (which depends on the cost function). Thus, the creation of bad candidate structures is not a problem.

The typical effect of mutate and crossover on the progress of the current best candidate structure using a GA is shown in Fig. 4a, whereas Fig. 4b shows the typical progress of the diversity which is defined as the number of bits along the simulated DNA where 0 and 1 both exist in the current population, and therefore has a value from 0 to 3 $m'N_i$ (where N_G=$2^{3m'}$). Notice that the diversity drops unless the mutation move class is used. By using schema theory [30, 31] (where candidates with certain sequences of 0's and 1's in a particular section are associated with a so-called schema) it can be shown how the GA can initially improve the quality of the candidate structures so rapidly [19]. Interestingly, if one of the bits within the simulated DNA of the entire current population does not correspond to that of the observed structure then it is impossible for the GA using only crossover as its move class to predict the observed structure. If, for example, the first bit for all current candidate structures were a 0 (like the two candidates in Fig. 3) and the observed has a 1, then a different move class would be required, like mutate.

Crossover works well when there is a large diversity in the current population. To assist in keeping a large diversity new random candidate structures can be included in the current population; simulating the arrival of immigrants [1]. The choice of move class is important. For example, consider the mutation process when trying to predict a structure using the minimum number of grid points possible ($N_G=N_i$) and where candidate structures that have ions on the same grid point have their cost function penalised heavily (ensuring that they are eliminated from the population). Doubling the number of grid points per variable would allow the possibility of moving ions half way between other ions. Alternatively a more appropriate move class (coordinated mutation) could be used where the coordinates of two ions within a current candidate are "switched" [1]. This type of move class has also been used in a SA approach [32]. As our simulations have periodic boundary conditions applied across the faces of the unit cell, a "translate" move class could be defined whereby all the ions are displaced by half or a quarter of a unit cell vector. Combined with crossover, this may help duplicate any correct local structure that is found in part of a current candidate (particularly helpful when the observed structure has a higher symmetry than that applied in the simulation). "Switching" has proved useful when the GA generated candidate structures with the wrong cation ordering, a common problem when cations are of similar size and charge [1]. This move class is also computationally faster to apply than the binary mutate move class operator as it does not require the conversion of the variables to/from DNA binary strings and the generation of a random number per bit rather than per ion. However, it is worth mentioning again that for crystal structure prediction the computing time spent on evaluating the candidate structures overshadows any time spent on generating candidate structures. When selecting which type of move class operators to use, small savings in the time required to generate each new candidate structure are not as important as minimising the number of candidate structures that are generated before the desired structure is found.

2.1.3
Systematic Search

There are alternative approaches to crystal structure prediction than that of evolutionary methods. These methods rationalise crystal structures by implementing ideas of geometric regularities in their atomic structure. One example is to apply the maximum-filling principle, or the ideas of closest packing of rigid spheres (for normal metals and their alloys, as well as for many ionic and molecular crystals). Using coordination numbers and the definition of structural units and their dimensionality have proved to be particularly useful for silicates and related materials. Details of these traditional approaches, which also include, for example, an explanation of continuous networks of glass materials, are exhaustively described in many crystallography texts [33, 34].

Here, I highlight recent work of systematic searches for topical catalytic materials, in particular, microporous materials such as zeolites. The frameworks of these structures are made up of corner-sharing tetrahedra blocks of SiO_4 or AlO_4 units which obey the Loewenstein rule [35] (empirical rule that between two AlO_4 units there must be at least one SiO_4 unit). The anions/cations are all two-/four-coordinated, respectively. The number of possible candidate structures can be reduced by what symmetry sites the ions may occupy for each space group, and how they are connected, in order to satisfy these coordination numbers. For example, if an anion is first placed on a site which has two mirror planes passing through it then, to generate a coordination number of 2, a cation must be placed on one of these mirror planes otherwise the anion will have a coordination number of 4. Generally, if the coordination of the cation is too large/correct/too small then reject/accept/continue by adding another anion, respectively. In a similar way, Treacy et al. [36] tabulated, for each space group and number of units within the system, the combination of symmetry sites the cations may sit on (so called T sites) and how these are connected in order to fulfil the coordination numbers of all ions within the reduced unit cell. Note that the anions are added later on the plane that is midway between two T sites (and with the observed T–O–T bond angle). Although more heavily constrained (compared with just fixing the content of the unit cell) this approach still requires one to locate the ionic positions that are, for example, on a symmetry axis or plane. Here, a global optimiser, for example SA, is used so as to produce a low value of a suitably defined cost function (least mean square deviations of interatomic distances and bond angles [37], say, see next section).

In a different approach, a mathematical scheme based upon combinatorial tiling theory [27] is used to generate structures made up of corner-sharing polyhedra. Again, a subset of possible candidate structures are produced which are then evaluated. In particular, a method for systematically enumerating crystalline networks made of tetrahedral building blocks has been developed [5, 38]. By defining the tetrahedral blocks to be those of SiO_4, AlO_4 or PO_4 units, the group of Bell, Foster, Klinowski and Simperler have generated hypothetical zeolite and related aluminium phosphate structures. An initial evaluation process is used to reduce further the number of candidate structures to examine more closely. The oxygen anions are placed midway between T sites, again such that there is a kink in the T–O–T bond angle. With reasonable initial bond distances, the lattice energy of the structures generated is minimised. For the stable structures, the lattice energy-framework density ratio is computed. The value of this ratio gives an estimate of how chemically feasible the particular hypothetical structure is. Their approach to isolating likely structures is based on earlier work by Price et al. [25], where hypothetical silicate frameworks were created by connecting sheets of tetrahedral units. In this method, different ways in which the T sites could be connected within sheets such that the coordination number of each T site was 3, are generated. Then, with the constraint that the T sites become four-coordinated, different hypothetical silicate frameworks are created by either stacking different combinations of these sheets or by stacking the sheets in a different way (displacing alternative sheets, say). An extensive review on hypothetical molecular sieve frameworks is given by Klinowski [26].

To evaluate the quality or goodness of fit for the candidate structures generated from either the various systematic or evolutionary approaches, a cost function is defined.

2.2
Evaluating Candidate Structures – Cost Function

2.2.1
Introduction

In order to assess the quality of the candidate structures that are generated, a cost function which measures the goodness-of-fit is defined and evaluated for each candidate. There are three popular methods of assessing the quality of a candidate structure: a direct comparison of the experimental and calculated powder diffractograms (R, say), the deviation of geometric parameters from that expected, and the cohesive, or lattice energy, of the candidate structures. Furthermore, the cost function can be a weighted combination of these or may be changed during the process of evaluating/predicting the candidate structures (see later, switching cost functions). The shape of the landscape mapped out by the cost function, as the unknown crystallographic coordinates and/or lattice parameters are varied, plays a fundamental part in determining whether or how easily the desired crystal structure will be found. For example, the surface created by R can be problematic, i.e. quite flat with a sharp minimum in one place [39]. Thus, when the current candidate structures have cost-function values only within the flat region there is a lack of helpful information on the whereabouts of the minimum of the observed structure. On the other hand, calculating the lattice energy using an ab initio electronic structure approach is generally too computationally expensive given the number of possible candidate structures to evaluate. An extensive review of determination, prediction and understanding of structures, using the energy landscapes of chemical systems is given by Schön and Jansen [40, 41]. What is required is a cost function that is a robust measure of geometry violation (provides sensible information even when the structure is completely wrong) and which is fast to evaluate. Attempts at this goal have been made by, as already mentioned, combining various aspects of the three traditional cost functions or by implementing a dynamic definition of the cost function. For example, using a combined "R and lattice energy" cost function [42],

$$\lambda E + (1 - \lambda)R, \tag{1}$$

where the weights are defined by the Pareto parameter λ, may eliminate minima that belong only to one part and may increase both the width and depth of the global minimum containing the desired candidate structure. This type of cost function has been used as a measure of quality for ionic and intermetallic compounds that were generated by a SA approach [43].

Before reviewing cost functions used in crystal structure prediction it is worth noting that good "book keeping" within a computer code can prevent many unnecessary calls to the cost-function subroutine which "evaluates candidate structure". For example, consider the rock salt system used to produce Fig. 4, with P_m set such that on average only one bit per new candidate structure is mutated. After 300 cycles of a GA, if all candidates are evaluated after each cycle there will be 30,100 calls to the cost-function subroutine. Even without "elitism" (copying the best candidate in the current population into the new population), by evaluating only candidates generated in the new population that were actually crossed or mutated (and copying a previously stored cost-function value otherwise) only around 19,000 calls to the cost-function subroutine would normally be required, leading to a possible 36% reduction in the computing time. A further reduction may be possible if the cost-function value for each different candidate structure evaluated is stored in a database that is checked as each candidate is generated (the unique DNA representation being used as the database pointer). Obviously there is a limit to the size of the database as the average time taken to search the database should be less than the time required to evaluate the cost function. Without restricting the size of the database, in the rock salt example, around 17,000 candidates were evaluated (and stored). In fact the number of candidates that were evaluated did not increase significantly if the size of the database was restricted such that the oldest 500 entries were deleted every time the database entries exceeded 1,000.

2.2.2
Cost-Function Definitions

The lattice energy based on the Born model of a crystal is still frequently used in simulations [14]. Applications include defect formation and migration in ionic solids [44,45], phase transitions [46,47] and, in particular, crystal structure prediction whether in a systematic way [38] or from a SA or GA approach [1,48]. For modelling closest-packed ionic structures with interatomic force fields, typically only the total lattice energy (per unit cell) created by the two body potential,

$$E = \sum_{i,j} \left\{ \frac{q_i q_j}{4\pi\varepsilon_0 r_{ij}} + A_{ij} \exp\left(- r_{ij}/\rho_{ij}\right) - \frac{C_{ij}}{r_{ij}^6} \right\}, \tag{2}$$

is required, where the summation over i (within the unit cell) and j (which includes translational images) creates unique pairs of atoms, a distance r_{ij} apart so as to avoid double-counting. Note that 3D periodic boundary conditions are applied to the unit cell and the Ewald summation [49, 50] is employed to compute the Coulomb term for the formal charges, q, on the ions. The remaining terms constitute the short-range Buckingham potential, where a radial cutoff between 10 and 25 Å is generally applied. The potential parameters, A_{ij}, ρ_{ij} and C_{ij}, may be derived using standard empirical fitting techniques [11]. Many published potential parameters can be found on a website database [51]. Although alternative functions can be used for computer simulations, the Buckingham form has proved to be particularly effective for ionic structures containing sensible bond lengths; however, when used as a cost function for structure prediction, problems arise when $r_{ij} \to 0$ as can be the case when new candidate structures are generated. The Lennard-Jones potential,

$$V_{ij} = \frac{A_{ij}}{r_{ij}^{12}} - \frac{C_{ij}}{r_{ij}^6}, \tag{3}$$

is a more robust functional form and has been used within SA by Schön and Jansen [52]. Additional terms are used when Eqs. (2) or (3) do not adequately model the system of interest. For example, if the transition-metal environment is distorted owing to the Jahn-Teller effect, then a nonspherical potential should be used [53]. Or, if polarisation effects on, say, the oxygen anions are important then the shell model [9, 54] is typically used. However, when assessing many candidate structures, Jahn-Teller and polarisation contributions to the energy could be added once the worst candidate structures have been discarded [55], see next section on switching cost functions. A three-body term [56] is commonly used when modelling silicate structures [57] where the O–Si–O bond angles are known to be around 109.47°, which corresponds to the perfect tetrahedral environment. If the stable polymorph at a particular pressure (where the unit cell dimensions are allowed to vary) and temperature is sought, then the Gibbs free energy should be used rather than the lattice energy.

For closest-packed oxides, a Pannetier-type cost function [58] is more robust and faster to evaluate than the lattice energy as defined earlier. Here, the bond valence model [59] is used to calculate the charge on the ions and the discrepancy with the expected value is used to measure the quality of the structure. With an additional term, the discrepancy in the expected and calculated coordination numbers, the cost function becomes

$$E = \sum_{i}^{\text{Unit Cell}} \left\{ k_b \left[|q_i| - \sum_j \exp B(R_{ij} - r_{ij}) \right]^2 + k_c [\text{Cn}_i - \text{Cn}_i']^2 \right\} , \tag{4}$$

where B and R_{ij} are known parameters defined by Brown [60, 61], Cn_i and Cn_i' are the calculated and expected coordination numbers of ion i. Different combinations of expected coordination numbers can be tried if the coordination numbers are not known. The weights, k_b and k_c, will obviously affect the outcome. For example, using this cost function and $k_\text{c}=0.0$, a GA successfully predicts the rock salt structure for BaO [1]. But by specifying coordination numbers of 4 (rather than 6) for the ions and increasing the weight k_c the caesium chloride structure can be generated. Note that changing either the size of the ions or the unit cell can have the same effect [1]. Another adaptation of the Pannetier cost function has the Coulomb repulsion between ions (with weight k_q) in order to help keep cations apart [1].

An alternative approach (global optimisation by basin-hopping [62]) is to use the minimised lattice energy as the cost function (Fig. 5). That is, rather than relaxing (to directly minimise the cost function) just the final structure in the SA approach or the better candidate structures from the GA approach (with a small population size), every candidate structure is relaxed as part of the evaluation process. With the "plateau style" landscape of the cost-function hypersurface (Fig. 5) the global minimum may be easier to locate. In planning which type of cost function to use this increased possibility of finding the global minimum must be traded against the computing time required to minimise every candidate structure. Combined with a SA approach this method has been designed by Wales and Doye [62] to predict the structure of clusters. However, although the coupling of local minimisation and GAs has been used to solve crystal structures [63], to my knowledge this type of cost function has not yet been fully explored for the prediction of inorganic crystal structures. In a GA approach, if all candi-

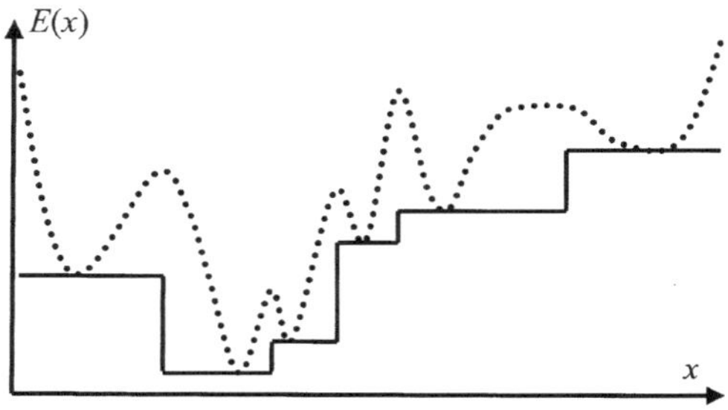

Fig. 5 1D example of a cost function hypersurface (*broken line*) and the landscape of the flat plateaux that is formed from the minimised cost function (*solid line*). Here, it may be easier to find the global minimum of the solid line rather than that of the broken line as the former contains no barriers

date structures are relaxed before being assessed, should the original or relaxed candidate structure be used for procreation?

The idea of just partially relaxing the candidate structures as part of the "procreation scheme" for new candidate structures could prove useful in later stages of a GA. For example, one can imagine newly generated candidates that had a new desired local structure but that also have ions so close together that the value of the cost function would make it unlikely to survive (chosen for future procreation). Before evaluating, it could be useful to "semirelax" (using the method of conjugate gradients [64], whereby the number of line searches [64] used to find the local minimum is restricted) the candidate structures first. That is, after an ion moves owing to the process of mutation, say, other ions are allowed to move slightly to make room for this ion and/or perhaps to fill the vacancy, where the ion originated. In the genetics analogy, the structures could be said to "mature" before being assessed. As already mentioned, the mature process would be useful when the population of structures already has cost-function values very near local minima (and the location of another local minimum is desired). For example, when the value of the cost function for a new candidate, before its structure is relaxed, is much higher than the current average value for the population but happens to be within the desired local well which contains the global minimum.

2.2.3
Switching Cost Functions

The strength of the GA is its ability to search and find wells in the landscape created by the cost function but not necessarily to find the totally relaxed structure associated with the local minimum within each well. That is, once a candidate structure is found whose value of the cost function lies within the well containing the desired local minimum, it is more efficient to relax the structure via a direct minimisation rather than waiting for a lucky combination of mutations. The latter is often why there is still a slow improvement of the best current value of the cost function, for the population, towards the end of a GA run (Fig. 4a

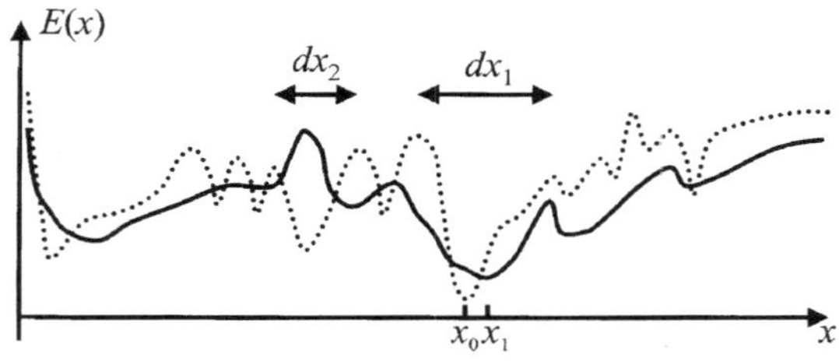

Fig. 6 1D landscapes of two fictitious cost functions, E. The desired structure has the cost-function value of $E(x_0)$

where the curve is very flat). Thus, the GA could be terminated much sooner if the process of generating the ideal structure includes relaxing the best candidate found. So the GA is used to find any structure that can be directly relaxed to generate the desired structure. For example, consider the lattice energy curve in Fig. 6. The GA (or SA) is used to find any point within the region dx_1, as $E(x)$ will then minimise to give the global minimum $E(x_0)$.

There is no reason why the same definition of the cost function should be used for both stages (GA and relaxation of the better candidates). In fact there is potentially much to gain from switching cost functions. Consider the two fictitious cost functions shown in Fig. 6, where the global minimum, $E(x_0)$, of the broken line corresponds to the desired/observed structure. Imagine trying to "design" a cost function, whose hypersurface is the solid line, that will be used in the GA stage. Of course, the point x_1 (corresponding to the global minimum of the solid line) must be within the region dx_1 [well of the broken curve containing $E(x_0)$] but x_1 need not necessarily equal x_0. With the fictitious cost functions shown in Fig. 6, if the GA found the structure with (or near) the global minimum of the solid curve, then by switching cost functions, this structure could be relaxed to the global minimum of the broken curve. There are two advantageous features of the designed solid curve when compared to the broken curve. First, the well containing the global minimum is wider, and so is easier to find (although a candidate structure should be found that has a value which is near the bottom of the well). Second, within the wrong region, dx_2, a local minimum has been replaced by a local maximum. Thus, the cost function (used in the GA stage) could be designed to bias the generation of structures away from unwanted minima in the lattice-energy surface (used in the second stage) and have a wider well than that of the energy well containing the observed structure (Fig. 6). For example, when trying to generate the NaCl rock salt structure (where the cubic unit cell dimensions and contents are fixed) the energy hypersurface will also have a local minimum which corresponds to the CsCl structure. As already mentioned, the discrepancy in the expected and calculated coordination numbers could be used to make the CsCl structure more unfavourable.

Now consider trying to generate the orthorhombic structure of $LaMnO_3$, where the corner-sharing MnO_6 octahedra are distorted (Jahn-Teller effect). To model these distortions would require including electronic effects in the cost function. However, the GA is used to generate an approximate structure which can be relaxed

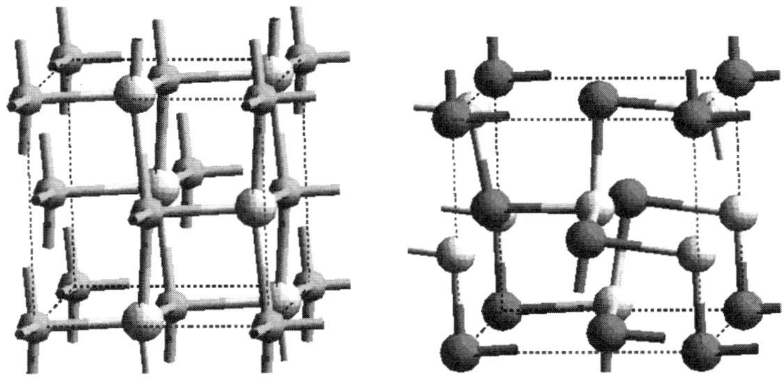

Fig. 7 Ionic positions of CdO and CoO, before lattice-energy minimisation was applied, for the best candidates that a GA produced after 50 cycles [1]. Reproduced by permission of The Royal Society of Chemistry

(after the cost function has been switched) to give the observed structure. Thus, the initial cost function can leave out electronic effects (which only perturb the structure) as it does not matter whether the better candidate structures generated by the GA contain distorted or symmetrical octahedra. Note also, when designing a cost function, the GA does not require the derivatives of the cost function.

To summarise, the idea is to use a "faster-to-evaluate", more robust definition of the cost function for the GA stage to generate approximate solutions. This approach is used by Woodley et al. [1], where the Pannetier-type cost function (Eq. 4) is used in the GA (stage 1) and the better candidates produced are (in stage 2) relaxed using the method of conjugate gradients to minimise the lattice energy (Eq. 2). Here, the lattice energy is considered to be a better measure of quality but is slower and less robust than the Pannetier-type cost function. Two structures produced from their GA are shown in Fig. 7. Both, by minimising the lattice energy, can be relaxed such that the ions move onto the more symmetrical rock salt lattice sites. However, if the lattice energy, described in Eq. (2), is used as the cost function, then the GA often fails because the lattice energy is badly defined when any of the interatomic distances approach zero.

3 Applications of Crystal Structure Prediction

3.1 Genetic Algorithms – Closest-Packed Oxides

3.1.1 *Multistage Scheme, Examples*

Consider the flow chart in Fig. 8, which portrays the steps within a GA used in the first stage, by Woodley et al., to predict the crystal structures of closest-packed oxides [1]. The initial population contains $2M$ random candidate structures created using the same unit cell dimensions and constituent ions (assumed to be

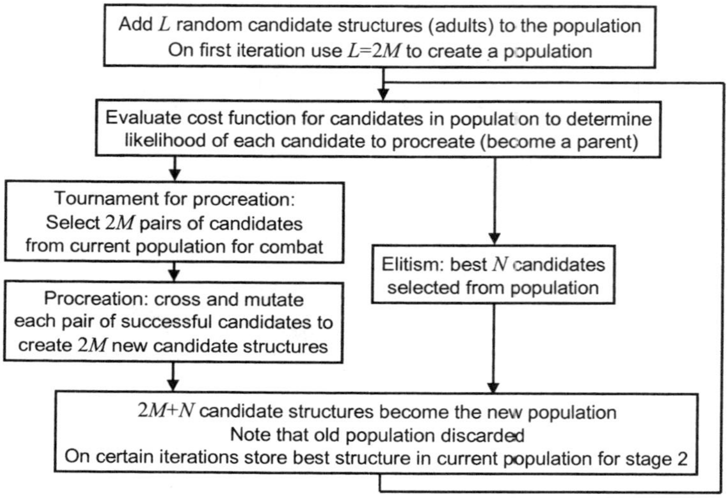

Fig. 8 Flow chart showing the various steps in a genetic algorithm used, by Woodley et al. [1], to predict known binary and ternary oxides

known and therefore fixed). The quality of each candidate is evaluated using a Pannetier-type cost function (Eq. 4). To introduce competition between the candidate structures, $2M$ pairs of candidates are then chosen at random for a "tournament" style selection procedure. Note, any candidate may be chosen more (or less) than once. For each pair, a random number between 0 and 1 is compared with a user-defined number P_t, the tournament probability. If the random number is smaller/bigger than P_t then the candidate with the lower/higher cost function is used to create the next population of candidate structures. The closer P_t is to 1.0 the more likely that the better candidate structures are selected (and chosen more than once). It is important that the diversity of the population of candidate structures does not diminish too quickly (where, in the extreme case, only the current best candidate survives). This is especially true in the earlier iterations of the GA, otherwise a global minimum of the cost function is unlikely to be found.

The move class operators crossover and mutate are then applied to these $2M$ parent candidates with probabilities P_c and P_m, respectively, to create $2M$ new candidate structures referred to as children. The old population of candidates, apart from the best N candidates (elitism), are then discarded (old age in the Darwinian analogy). The $2M+N$ candidates together with L more random candidate structures ("foreign" structures migrating into the population) form the new population for the next iteration, which begins with the quality assessment of this new population. The best candidates generated from the GA after 50, 100, 150,...300 iterations are saved for the second stage, where the lattice energy (Eq. 2) for each candidate structure is minimised. As the superior candidates prosper and the weaker candidates perish, children emerge that, after relaxation in stage 2, resemble the observed or desired crystal structure.

Woodley et al. implemented this method within the GULP code [11, 12], whereby the better candidate structures were relaxed in a second stage after switching cost functions (see earlier). They defined a default set of parameters (M=50, N=L=2, P_t=0.9, P_c=0.5) for their GA such that their two-stage approach could be used to generate over 30 known binary oxides and various known ternary oxides which included the perovskite, pyrochlore and spinel structures [1]. The value of P_m was set such that on average only one bit per new candidate structure is mutated. The mesh used across the unit cell initially contained 16^3 grid points. N_G would then be increased by a factor of 2 after every ten iterations to a maximum of 264^3 grid points. In stage 1, input data included the fixed unit cell dimensions and constituent ions for known closest-packed structures. The use of a coarse grid across the unit cell during the earlier iterations is beneficial, as better candidate structures with evenly spaced ions are quicker to generate. Moreover, searching this smaller set of different possible candidate structures may help the GA to find the region of the cost function hypersurface that contains the global minimum before the diversity falls as one candidate structure begins to dominate. Note that using a finer mesh a better candidate can gradually improve its bond distances through small mutations and thus obtain a value of the cost function near a local minimum. If this value is much lower than the average value of the cost function for the current population then this candidate may begin to dominate (diversity drops) before the GA has had a chance to search for other, possibly lower, local minima.

This multistage scheme (GA with cost function based on Eq. 4 followed by a relaxation of the better candidates with lattice energy as the cost function) was previously implemented by Bush et al. [2], who developed the code GASPP [65] for stage 1. Although the GA parameters are similar (P_c=0.8, P_m=0.003, L=0), there are differences: (1) the grid across the unit cell is fixed; (2) elitism, with N=2, is used to expand the initial population from M=10 to 50, whereafter the worst two candidate structures of the new population are replaced by the best two candidates of the old population; (3) a more sophisticated way of choosing candidate structures for parenthood (procreation) is employed. Here, the selection of 2M candidates is statistically weighted using the cost-function values of the candidates such that the better candidates are exponentially more likely to be chosen (zero probability is given to candidates with a cost-function value greater than a defined threshold). Bush's cost function integrates the use of the average observed coordination numbers [60] and the more generalised use of bond valences [61]. For example, for each ion within the unit cell, the bond valences are added to the magnitude of its charge unless the neighbouring ion is (of the correct type and) within its first coordination shell.

This approach, combined with a final refinement in a third stage using the Rietveld program GSAS, was successfully applied to the previously unsolved structure of Li_3RuO_4; the structure is shown in Fig. 9b. Initial cell dimensions, obtained from the X-ray powder diffraction profile, and the contents (six Li^+, two Ru^{5+}, eight O^{2-}) were used as input parameters for stage 1. The better candidate structures, produced using the GASPP program, were selected after 1,500 cycles. With the cell dimensions still fixed, in stage 2, these structures were relaxed (lattice energy minimised) using the GULP program [12]. The simulated X-ray

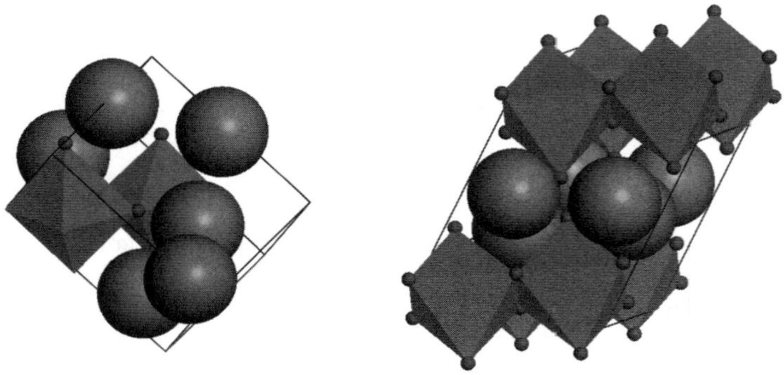

Fig. 9 The crystal structure of Li_3RuO_4 after stage 2 (*left*) (before the Li^+ coordinates were refined) and after stage 3 (*right*) (the predicted crystal structure) where the *large* and *small spheres* are the Li^+ and O^{2-} ions, respectively, and the Ru^{5+} are within ruthenium-oxygen polyhedra

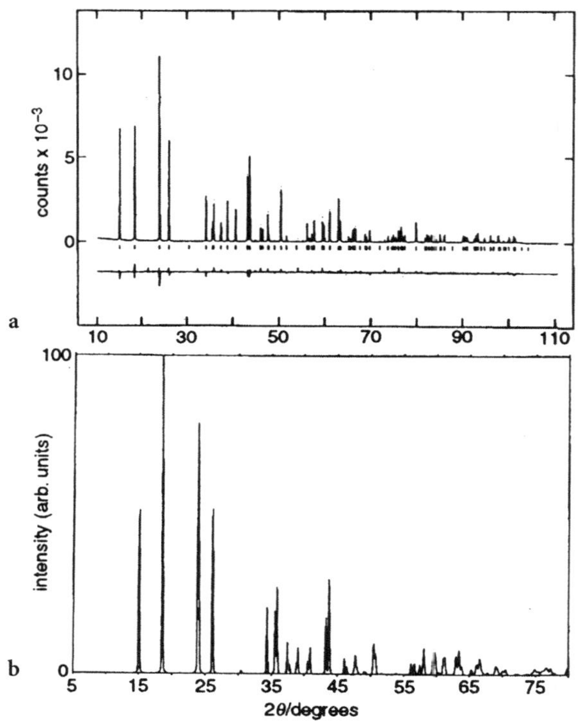

Fig. 10 **a** Observed, calculated and difference X-ray powder diffraction profiles of Li_3RuO_4. **b** Theoretical X-ray powder pattern calculated using the structure shown in Fig. 9a [2]. Reproduced by permission of The Royal Society of Chemistry

diffraction profile, Fig. 10b, of one of these relaxed candidates showed remarkable agreement to that observed. At this stage, upon looking at the sublattice formed by the ruthenium and oxygen ion positions, Bush et al. were able to assign space group *P2/a* to the structure. The lithium positions were then refined (as well as the cell dimensions) using the Rietveld program GSAS [23]: the observed, calculated and difference X-ray powder diffraction patterns are shown in Fig. 10a.

I note that the cost function and GA parameters used in the approaches just described, and in general (even within the same research group), are continuously modified to suit the particular system(s) of interest.

3.1.2
Polymorphism

With the constraint of the same chemical formula, there can exist a number of (meta)stable structures, the phenomena known in crystallography as polymorphism. The global minimum of the ideal cost function might not be the only important minimum of interest. Under the same physical conditions more than one polymorph may exist in the same crystalline sample. Mathematically this happens to be when the local minima and the global minimum are of a similar value. One can either search for the most stable structure under given conditions or try to identify all important polymorphs. Incidently, α-quartz has two equally valid global minima where the structures are mirror images of each other (enantiomers).

The cost function can be chosen to fix the physical conditions and examples are presented later. However in this section I will continue to assume that the unit cell dimensions are fixed (as they are obtained from the X-ray powder diffraction profile). For different fixed cell dimensions, different physical conditions are "simulated"; therefore, we can expect different polymorphs to be generated by the GA.

The approach of Woodley et al., described earlier, has been used to generate three of the different polymorphs of TiO_2 (rutile, anatase and brookite) [1]. The interatomic potentials used for the three polymorphs of TiO_2 and the unit cell dimensions (and contents) were taken from Freeman et al. [66], who successfully generated these polymorphs using a SA approach (see later). Note that the discrepancy between the best generated ionic coordinates (Fig. 11b) is discussed in Ref. [1] and is ignored here as this generated structure can easily be corrected using an X-ray refinement package in a "final stage".

Of the three polymorphs, the rutile structure was the easiest to generate. Moreover, two other polymorphs (see Fig. 11), one when searching for the brookite polymorph and the other when searching for the anatase polymorph, were also readily found when using different settings in the GA. The former, shown in Fig. 11a, has the structural topology of TiO_2 (II) [67] and a similar lattice energy but a higher cost function than the brookite polymorph and initially dominates the population after stage 1 (number of iterations is less than 250). The latter polymorph (in Fig. 11b), the crystal structure of Cu_2O stretched along the *c*-axis, has a lower cost function than anatase. So, if the best candidate structures after 600 iterations were examined only the stretched Cu_2O polymorph would appear.

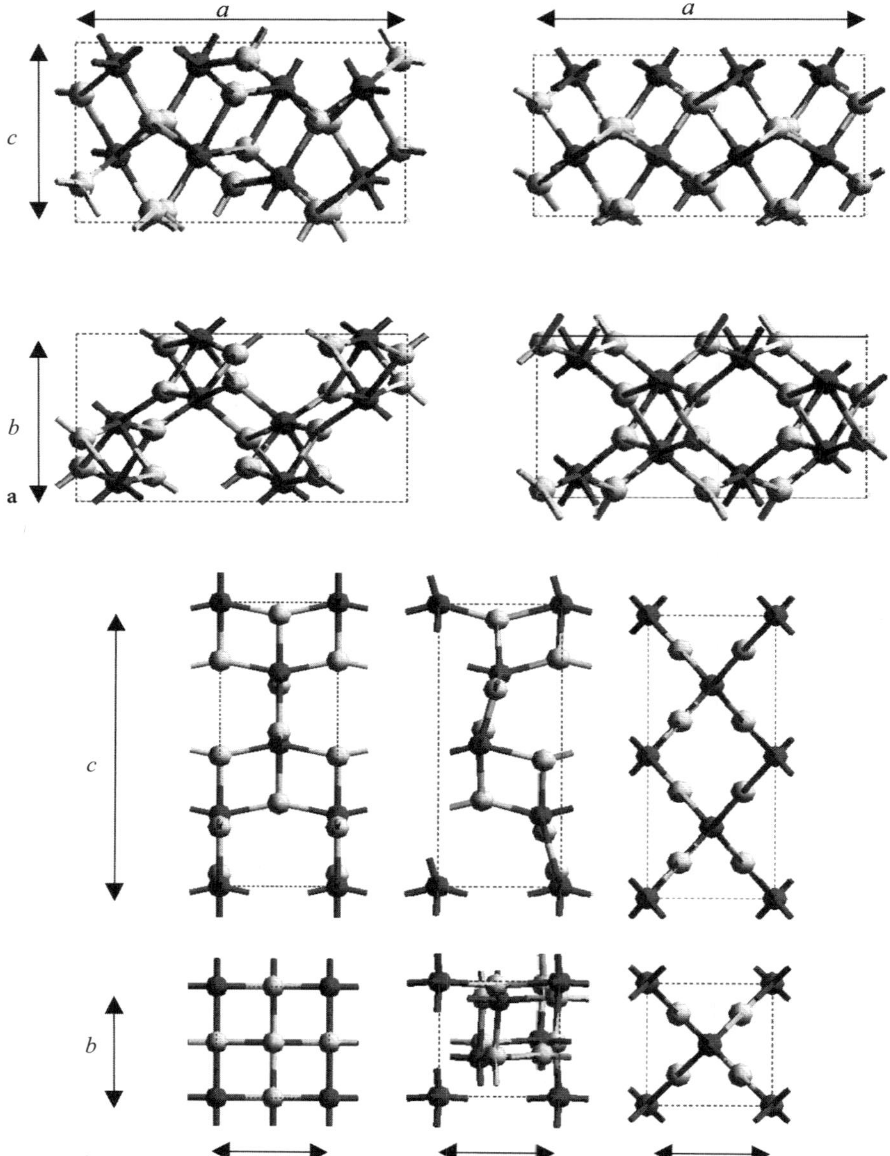

Fig. 11a,b Five different TiO_2 polymorphs viewed in two perpendicular views [1]. **a** The observed/generated structure of brookite and another polymorph after stage 2 which have cost-function values of 49.4 and 52.4 and lattice energies of −112.3 and −112.4 eV per TiO_2 unit, respectively. Both candidates have the cell dimensions and constituents of TiO_2 brookite. **b** The observed anatase structure, the best candidate and another readily found polymorph after stage 2 which have cost-function values of 41.6, 22.6 and 12.5 and lattice energies of −112.10, −112.16 and −112.11 eV per TiO_2 unit respectively. All three candidates have the cell dimensions and constituents of TiO_2 anatase. Reproduced by permission of The Royal Society of Chemistry

It is thus important to realise that the candidate structure with the lowest cost function is not necessarily the crystal structure that is observed. Furthermore, there is a need to save (for further analysis in stage 2) the best candidate after every 50 cycles, say, rather than just the best candidate structure at the end of stage 1. Ultimately, a comparison of the simulated and observed powder diffractograms is required in a "final stage". It is also noted that, after stage 2, the anatase structure had a lower lattice energy (a better measure of quality) than the stretched Cu_2O polymorph.

In practice the cost-function weights may also be varied, and as the anions in the anatase polymorph have a different coordination number from that observed in the Cu_2O polymorph then increasing the weight k_c could make the anatase polymorph more likely to be generated. It is, of course, possible that unexpected structures generated by the GA could be prepared by using alternative synthetic methods or be found in nature under nonambient conditions. The generation of the TiO_2 polymorphs has shown that where large unit cells are used there is an increased possibility of generating other dominant polymorphs. This happens when the unit cell can accommodate the replication of smaller unit cells required by another polymorph (not necessarily that of TiO_2), for example, the unit cell dimensions of anatase and brookite were 1×1×2 and 2×1×1 supercells of the stretched Cu_2O and TiO_2 (II) polymorphs.

3.2 Simulated Annealing – Closest-Packed Structures

3.2.1 *Fixed Unit Cell Parameters*

GAs are not the only method to be employed that can generate more than one of the TiO_2 polymorphs from the knowledge of the unit cell dimensions and its constituent ions. Both Freeman and coworkers [48, 66] and Reinaudi et al. [68] have implemented a SA technique where the experimentally observed ionic coordinates are successfully obtained, starting from random initial coordinates. The cost function employed by Freeman and coworkers during their first stage is based on the simple two-body r^{-12} interaction potential, weighted such that the repulsion between anion-anion and cation-cation pairs is twice as strong as that felt between cation-anion pairs. In their earlier publication they reported the success of their approach in generating the rutile polymorph. Here a Monte Carlo driven SA was used to produce candidate structures that, although the ions were initially randomly positioned, had sensible bond distances. In a second stage the lattice energy based on the Born model (including shells, which allow for the electronic polarisation) was used as the measure of quality (Eq. 2) when the candidate structures were quenched (relaxed, but with the lattice parameters still fixed). Note that for sensible candidate structures, the lattice energy is a better measure of quality than the cost function used in stage 1. This is particularly true if the interatomic distances are similar to that of the structure used when refining the interatomic potential parameters. In a third stage, after the candidate structures were relaxed under constant pressure, the particle density is calculated (which can be compared to that observed). Applied to 50 different random candidate structures (unit cell containing two formula units of TiO_2), the METAPOCS code [69] generated the rutile structure 41 times [48]. The other nine candidates had higher lattice energies and lower calculated densities (unit cell volume expanded during stage 3).

Upon further application of their technique to the various polymorphs of TiO_2, the method was refined [66]. The initial random structure had the additional constraint that significant steric overlap of ions was not allowed (new ionic coordinates were continuously generated until a suitable random site, not too close to any ion already randomly placed, was found). After the SA stage, where a Metropolis Monte Carlo scheme is employed, the candidate structures were quenched with respect to the r^{-12} cost function and then with respect to the lattice energy under constant volume, whereas previously the lattice energy was minimised (quenched) under constant volume first, then under constant pressure. They pointed out that the change in the topology of the crystal structures does not change significantly when the unit cell parameters are relaxed. A further reduction of the degrees of freedom is achieved by eliminating the shell model in the definition of the lattice energy. Again, they found no significant structural changes when the shell model was removed. The rigid-core representation of the ions is also used in the GA method discussed earlier [1].

Using the observed unit cell and contents for the respective polymorphs, they were able to generated the rutile (43 out of 50 attempts), anatase (four out of 50 attempts) and brookite (30 out of 50 attempts) polymorphs using their refined multistage SA approach. Thus, the highest rate of success was found for the rutile polymorph, which has the smallest number of degrees of freedom (number of ions is half that of anatase and a third that of brookite). The other candidate structures generated using the rutile and anatase unit cell and contents, respectively, had higher lattice energies; however, a different polymorph, 2×1×1 supercell structure of TiO_2 (II) [67], with a lower lattice energy was found (20 out of 50 attempts) when using the observed brookite parameters. The structural topology of TiO_2 (II), as shown in Fig. 11a, was also found when a GA approach was later applied to the same system [1]. Whether it is desirable or undesirable for a crystal structure prediction package to produce other polymorphs is up for discussion. Importantly, being able to compare the observed and simulated X-ray powder diffraction pattern, the brookite polymorph need only be found once.

A similar success rate, four and 44 out of 50 attempts for the anatase and rutile polymorphs, respectively, was more recently found by Reinaudi et al. [68]. Their initial work differs from that of Freeman and coworkers in that the lattice energy (Eq. 2 with identical potential parameters) rather than the r^{-12} cost function was used as the measure of quality during the SA stage. Although a better measure of quality for candidate structures that are similar to the desired structure, this definition of the cost function can be problematic, however, if there are very short interatomic distances. Thus, the benefit of using the r^{-12} cost function, to help prevent creating candidate structures that may contain collapsed regions, is lost.

More importantly, they showed that the rate of success in finding the anatase polymorph can be dramatically increased if symmetry constraints are imposed [68]. For example, by imposing I centring or a 4_1 screw axis the correct anatase

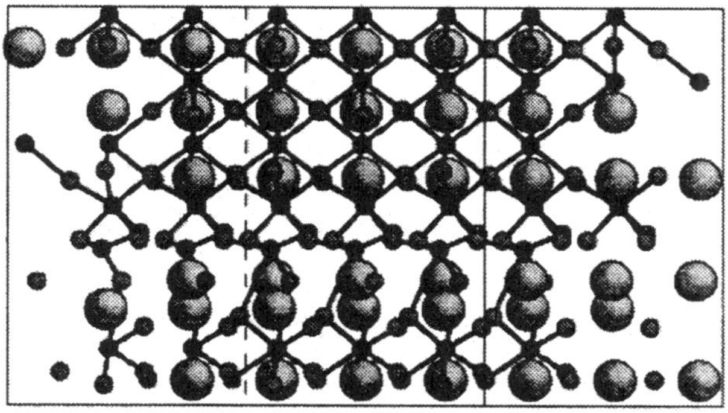

a

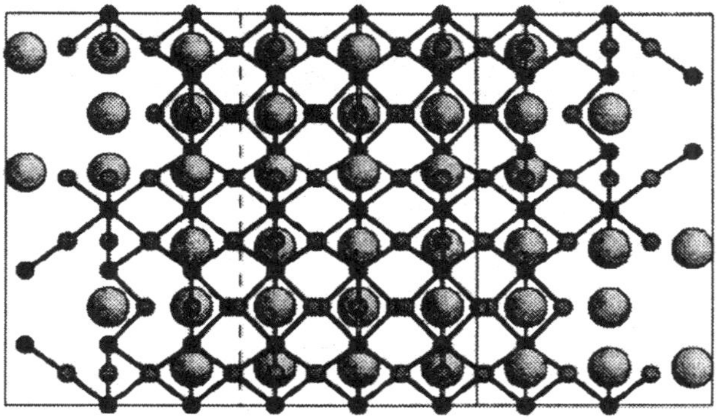

b

Fig. 12 a Lowest-energy structure obtained for $BaTiO_3$ with no symmetry imposition and **b** the crystallographic structure of $BaTiO_3$ viewed along the (210) direction. Although the structure in **a** is not the correct one, a large portion of the unit cell resembles that of the real structure [70]. Reproduced by permission of The Royal Society of Chemistry

structure was found 30 or 37 times out of 50 attempts, respectively. Note that the number of degrees of freedom is also reduced: when I centring is imposed there are six rather than 12 independent ions and when a 4_1 screw axis is imposed there are only three independent ions (one of which can be arbitrarily fixed to stop translations). The incorporation of space-symmetry elements, that are obtainable from the powder diffraction data, can also increase the accuracy of the atomic positions.

Their approach was then successfully applied to the generation of the ternary crystal structures $A\mathrm{TiO}_3$ [70]: cubic $SrTiO_3$ (found 40 out of 40 attempts), orthorhombic $CaTiO_3$ (found 27 out of 40 attempts) and hexagonal $BaTiO_3$ (found 18 out of 40 attempts). Here, the imposition of symmetry elements was not required for the cubic structure and a glide plane was imposed for the orthorhombic and hexagonal structures, without which the success rate was 27 and zero out of 40 attempts, respectively. Any criticism about whether improving the success rate is necessary, as the desired structure need only be found once, is removed, as without the improvement the hexagonal structure of $BaTiO_3$ was not generated. It can be seen in Fig. 12 that, without the imposed glide plane, the correct topological structure is only found in part of the cell. Thus, the global optimisation had problems try-

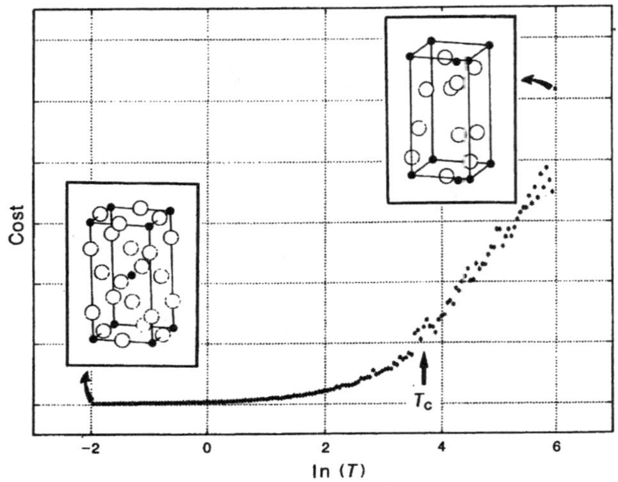

Fig. 13 Cost function as a function of the logarithm of temperature during the annealing of NbF_4. *Inset*: structural arrangement of NbF_4 at the beginning of the simulation (*right*) and as obtained by simulated annealing (*SA*) (*left*) [58]. Reproduced with the kind permission of the Nature Publishing Group (http://www.nature.com/)

ing to replicate this throughout the cell (an increasing problem for unit cells containing a larger number of ions as grains can form). Without imposing symmetry elements cation disorder was found in the other candidate structures generated for $CaTiO_3$: a problem met by Woodley et al. [1] in their GA results especially where cations with the same charge have similar ionic radii.

Reinaudi et al. only include centring, screw axes and glide planes, that are obtainable from the powder diffraction data, to avoid the problems associated with certain symmetry elements. For example, an ion that approaches a mirror plane will be repelled by its image; so a new move class would be required in order to allow ions on and off mirror planes. They also tried a refined cost function (initially the lattice energy), such that it included an additional additive term that required the calculation of the simulated structure factors [21]. Here, they took advantage of the fact that certain structure factors should be identical; when trying to generate the anatase polymorph, any differences result in the lattice energy being penalised by up to 20.0 eV per type of ion and set of equivalent reflections. The use of their new cost function improved the success rate of generating the rutile and anatase structures from 88 and 60% to 100 and 72%, respectively.

It should be noted that the structure of NbF_4 was the first successful ionic structure to be generated using a SA approach, where the unit cell dimensions and contents were fixed [58]. The cost function used contained two weighted contributions, a Coulomb part and a discrepancy term, which was based upon the bond valence model; see discussion of Eq. (4). The initial and predicted structures for NbF_4, as well as the value of the cost function during the SA stage, are shown in Fig. 13. This method was later used to solve the structure of the first dirutile compound, $LiCoF_4$ [71].

3.2.2 *Unit Cell and Coordinates Unknown*

So far the emphasis has been on solving the crystal structure from the knowledge of the unit cell and ionic content. The motivation for this work is to provide an automated procedure to help determine or solve the crystal structure of new compounds that are synthesised in a powder form. Of course the methods developed can generate other structural topologies and perhaps a new, yet to be synthesised, crystal structure. However, the task has been to solve a particular structure and so one might extract more information from the experimental data to aid the prediction process (e.g. use of symmetry elements). Thus, the number of unwanted possible (meta)stable structures, or polymorphs, that could be generated may be reduced. In this section, the emphasis is on finding all the important polymorphs for a particular chemical formula.

In this purer, or general, view of crystal structure prediction Schön and Jansen [52] have allowed the unit cell to assume an arbitrary shape and volume. This increase in the number of degrees of freedom will increase the number of different (meta)stable polymorphs that can be generated (and unfortunately significantly increase the number of possible undesired candidate structures). They implemented a standard stochastic SA algorithm [72] where a number of Monte Carlo steps are performed (change in atom positions and unit cell size) per value of the temperature parameter, T, which is decreased according to the rule $T_n=f^n T_0$. The initial systems of interest were the crystal structures formed by pure and mixed noble gases [52, 73] starting from a random configuration of atoms in a large cubic unit cell where a small nonzero external pressure was applied during the simulations. The enthalpy (per atom) was used as the cost function. Here, the sum of Lennard-Jones interactions, Eq. (3), was chosen and was added to the product of the pressure and volume. Two extra move classes were tried, both of which modify the unit cell dimensions: (1) the cell parameters were changed whilst the fractional coordinates of the atoms/ions were kept fixed (thus, a uniform increase in unit cell volume causes an increase in all interatomic distances); (2) the cell parameters were changed by slicing off a section of the unit cell or by adding on a section unit cell (interatomic distances not squashed nor stretched). The number of atom moves required to allow the system to settle after a volume change via slicing or adding to the unit cell was about 5 times less than the number of atoms within the unit cell. Thus, the latter move class could be applied more frequently and also proved more successful in generating structures that did not contain defects (distortions or holes in the closest-packed structures generated).

This technique was then successfully applied to simple ionic compounds [74]. As the individual constituents of the unit cell were now charged, they modified their cost function to include the energy contribution from the Coulomb interactions between the point charges (which could also be subject to change during the SA stage [75]) centred on the ions and the sum of the respective ionisation energies and electron affinities. To help the cost-function convergence, with respect to radial cutoff, the Coulomb term was damped exponentially (within their simulations they showed that the degree of damping did not significantly affect

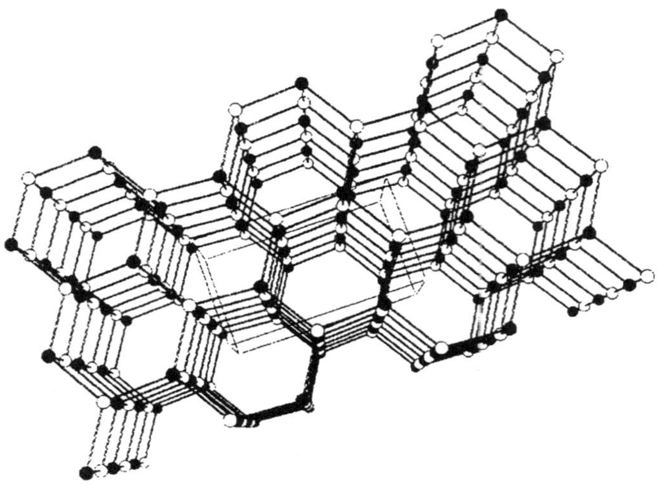

Fig. 14 A structure generated by a SA approach, with fivefold coordination of Na by Cl (Cl-trigonal bipyramids around Na). *Black circles* indicate Cl^- ions and *white circles* indicate Na^+ ions. Reprinted from [74], with permission from Elsevier

the structures generated). Adding the chemical potential (such that the cost function is now based on the free energy), the number of atoms/ions was also allowed to change during the simulations [75]. The Lennard-Jones potential parameters were scaled using the sum of the ionic radii. The ratio between applying the ion and volume change move class (length and direction of the cell vectors) operators was 4:1. Long SA runs were implemented to find the candidate structure with the global minimum of the defined cost function. For the binary systems considered, they found either the observed structure for the compound or a sensible alternative polymorph (e.g. CsCl structure for the observed NaCl structure of CaO).

The Lennard-Jones potential parameters affect the size and hardness of the ions as well as the strength of the van der Waals interactions. There is no guarantee of the "transferability" of the potential parameters employed within the cost function. Interatomic potential parameters previously fitted such that a minimum of the cost function corresponds to a known structure may not create a minimum in the cost function for the desired unknown structure. Thus, when shorter SA runs were employed to find metastable structures (higher-energy local minima candidate structures) of NaCl, the Lennard-Jones potential parameters were changed (within a sensible range). Note that, as discussed earlier, a better cost function can always be used on the candidate structures generated. A whole range of metastable candidate structures were found; some not previously observed, as in the example shown in Fig. 14. The kinetic stability of these metastable candidate structures was examined by finding the relative height of the lowest energy barrier (saddle point) to another lower local minimum on the cost-function hypersurface. They did this using the energy-lid method [76, 77], where several random distortions of the metastable structure are made (with the restriction that the value of the cost function did not increase beyond an upper limit, or threshold) before the candidate structure is quenched. Several runs are made per value of the threshold, which is gradually increased. Repeating this process, the cost-function hypersurface is explored; the height of the energy barrier is the difference between the cost-function value for the original metastable structure and the value of the threshold, or lid, when upon quenching a different structure is found for the first time.

Using a larger number of ions in their simulations they came across the multidomain problem discussed earlier whereby two different polymorphs begin to grow within different regions of the unit cell. However, their solution is rather unique in that they use the structural information found within each of the domains as a separate starting point in further simulations. The program developed for searching for such domains is called Findcell [78].

For a more complete review of the earlier work of Schön and Jansen (which includes the generation of closest-packed ternary systems, the use of the Ewald summation for the Coulomb contribution to the cost function [49, 50], and Hartree-Fock calculations [79] for the better candidate structures generated) see Ref. [75].

As mentioned earlier, Schön and Jansen are interested in generating, or finding, all the important (meta)stable candidate structures and to provide information on their stability (and eventually guidance on possible synthetic routes). Using this approach the structures generated have included not-yet-synthesised compounds, such as Na_3N [3]. Further work includes the systems AB_2 (A=Mg, Ca, B=F, Cl) [4], Mg_2OF_2 (see Fig. 15, where another example of an interesting yet-to-be-synthesised structure, generated by their approach, is shown) [32] and high-

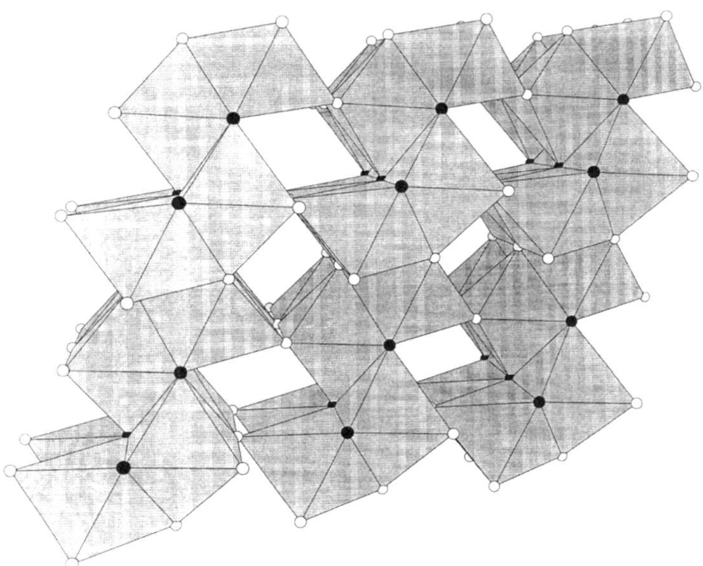

Fig. 15 Possible structure of Mg_2OF_2. *White* and *black circles* represent the F and O atoms, respectively. The *grey octahedra* contain Mg. Reprinted from [32], with permission from Elsevier

pressure phases of $(M,M')_3N$ and M_2S (M=Li, Na) [80]. During the simulations for each of these systems, a different definition of the cost function was employed. For example, in the AB_2 systems the ionisation, chemical potential and pressure terms were dropped (such that the cost function is the sum of two contributions: the Lennard-Jones short-range interactions and the Coulomb interactions which are calculated using the Ewald summation in the formulation of De Leeuw et al. [81]), whereas the pressure term was kept for the $(M,M')_3N$ and M_2S systems and the ionisation term for the Mg_2OF_2 system. The move class operators used within the SA algorithm were also refined. For example, in AB_2 the "change the ionic charge" was removed, "swap/exchange ions" was added and both types of move class operators for changing the unit cell dimensions were employed, although in each system random displacement of ions was still the most frequently used move class (typically around 75%). It is clear that there is no ideal cost function or set of move class operators that is suitable for all systems and that some degree of experimenting is advisable.

3.3 Framework Crystal Structures

3.3.1 *Genetic Algorithm Approach*

Consider the multistage process, described earlier, of Woodley et al. [1]. In the quenching (second) stage, the lattice energy is used as the measure of quality. It seems reasonable that this stage is still appropriate for microporous or mesoporous materials as, for example, the refinement of MeALPO-36 by minimising the lattice energy has already proved useful in solving framework materials [82]. However, in the search for a candidate structure that when quenched will relax to the desired framework, the lattice energy may favour clusters of closest-packed ions within the unit cells as the framework structures are metastable (and the closest-packed form is often the stable polymorph). Trying to encourage the GA to generate framework structures from the starting point of randomised ionic coordinates is one task, getting the right minimum on the cost-function hypersurface is another (there are many ways to buckle the framework to produce different metastable candidate structures).

In the last 3 years, Woodley et al. have employed "exclusion zones" (EZs) within their GA (used in stage 1) in order to bias the generation of framework candidate structures [18, 55, 83]. During stage 1, atoms/ions are encouraged to stay out of the EZ, which are regions within the unit cell; Fig. 16. In their implementation, the EZ can be either ellipsoidal (like the spherical EZ in Fig. 16a for sodalite), elliptical cylinders (channels, Fig. 16b) or planar (EZ can contain known parts of layered crystal structures, say). These were employed using two different methods; directly by using a grid where no points were within the EZ or indirectly by adapting the definition of the cost function. In the indirect method, when an ion is found within the EZ, the candidate structure is heavily penalised such that it is unlikely to be selected for procreation (and so will vanish from the population). Variations to this method could include a soft boundary whereby the penalty on the cost function is weighted with respect to how deep into the EZ the ion is. As the cost function is generally expensive to compute, they do not compute the proper cost function for a candidate structure if one of its ions is found within the EZ. Note that, when the population contains only structures which have at least one ion within the EZ, it is important that the value of the cost function for $n+1$ ions found within the EZ (images outside the unit cell are not considered) is greater than those with n ions found within the EZ. In the direct method a linear grid is first defined over the unit cell. The grid points within the EZ are effectively removed by mapping the binary representation to the grid points outside the EZ. The advantage of this method, rather than just squashing the grid points between the EZ, is the "uniform" sampling of the unit cell and the reduced length (number of 0's and 1's) of the binary representation (cost-function hypersurface reduced). The disadvantage is that some grid points may be represented twice (as it is not always possible to have 2^m grid points outside the EZ) and the mapping between fractional coordinates and the binary representation is more expensive. Note that when the grid points within the EZ are not excluded,

$$f = \frac{1}{2^{m'}-1} i \tag{5}$$

where f is the fractional coordinate for an ion on grid point i (decimal integer of the binary representation for the particular ionic coordinate) and $2^{m'}$ is the number of grid points along the length of the unit cell.

Whether the direct or indirect approach to EZ is more efficient will depend on the volume of the EZ compared to the unit cell volume [83]. For the indirect approach, an increase in the EZ volume will result in more candidate structures being generated with ions within the EZ, whereas for the direct approach both the number of different possible candidate structures and the effort required to compute the fractional coordinates from the grid points are reduced. For the sodalite example in Fig. 16a, if an accuracy of the ionic coordinates of less than 0.1 Å is required after stage 1 then a 128^3 grid would be required if no EZ were used. Using spherical EZs, as shown in Fig. 16a, of radius 4.0 Å, leaves just 500,778 grid points (a reduction of 2 bits within the sequence of 0's and 1's per unknown), whereas for a radius of 4.5 Å only 8,796 grid points are left. Beware, when employing EZs, particularly for the indirect approach, the chosen move class operators must ensure that the desired structure is still accessible.

With the direct approach to EZs, Woodley et al. generated the silica form of zeolite JBW (Fig. 16b) from an initial population of candidate structures which had random ionic coordinates, within the full unit cell, for both the Si^{4+} and the O^{2-} species [83]. The channel EZs are shown in Fig. 16b; the existence of which is assumed to be detectable via surface images and sorption experiments.

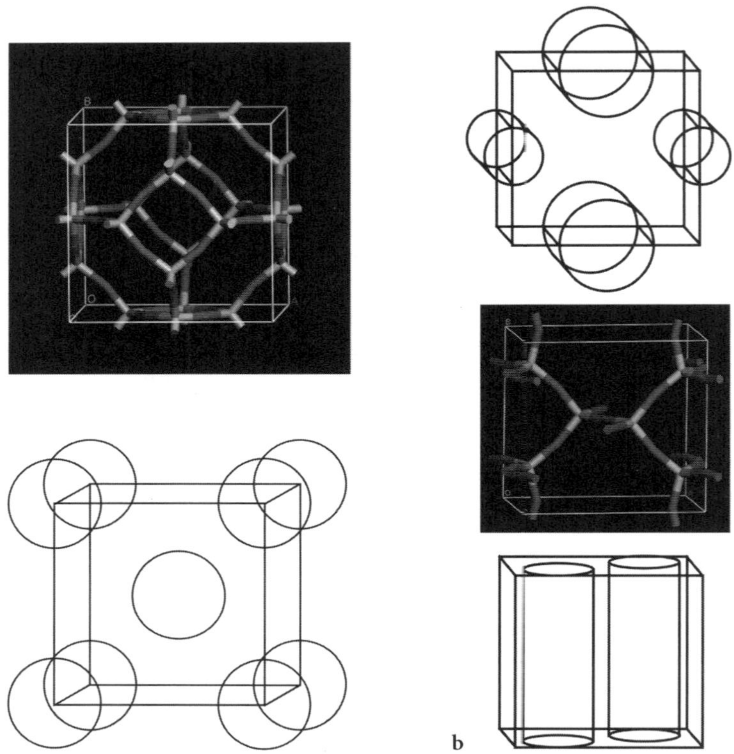

Fig. 16 Possible spherical and cylindrical exclusion zones (regions of the unit cell containing no ions) for the structures of SiO_2 **a** sodalite and **b** zeolite JBW

3.3.2
Simulated Annealing Approach

A SA approach to generating framework materials was successfully applied by Deem and coworkers [84–86], Akpoiaye et al. [87] and more recently by Boisen and coworkers [88–90]. Ignoring the differences in the definitions of the cost functions, the different approaches used can be divided into two groups depending on whether the SA is used to find only the cation, or T, sites or all ion sites. For example, Boisen and coworkers have generated a vast number of candidate framework structures [88, 90] using the geometry and a potential-energy function derived from fitting to the disilic acid molecule and starting from a randomised set of n silicon and $2n$ oxygen coordinates (n=3, 4, 6) within the full unit cell. In contrast, in the approach of Deem and Newsam [84, 85], the constituent oxygens are added after the locations of the T sites are found.

Consider four-connected framework structures that are made up of corner-sharing TO_4 tetrahedra units, where T represents Al or Si species. In the SA approach employed by Deem and Newsam [84] the coordinates of the T ions are

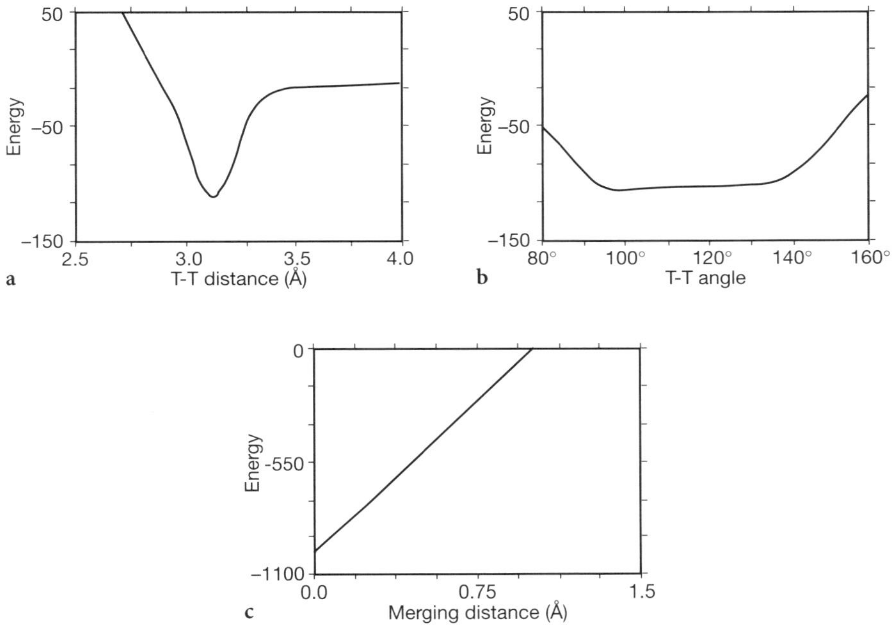

Fig. 17 Energy as a function of **a** *T–T* distance and **b** *T–T–T* angle used in the simulation procedure (calculated as smoothing spline fits to Boltzmann equilibrium interpretations of the histogrammed data taken from 32 representative zeolite crystal structures). Only the central portions are shown. **c** The contribution to the energy sum for the merging of two symmetry-related atoms; merging is only permitted when the two atoms are at less than a defined minimum distance [84]. Reproduced with the kind permission of the Nature Publishing Group (http://www.nature.com/)

found, corresponding to a local minimum of their cost function which is based on *T–T* distances, *T–T–T* angles and an additive penalty of, typically, 300, 600, 800 and 1,000 for a discrepancy in the number of nearest neighbours of 1, 2, 3 and 4, respectively (where four *T* ions are expected within 3.7 Å). The functional forms of the other terms are shown in Fig. 17. A distance least-squares approach, based on *T–T*, *T*–O and O–O interatomic distances [37, 91], is then used once the oxygen species have been added at the midpoints between pairs of *T* species. This method helped determine the previously unknown structure of lithium gallosilicate. Deem and Newsam [85] later refined their method by including the target powder diffraction pattern and were able to generate over 5,000 hypothetical structures. Akpoiaye et al. [87] used the method of Deem and Newsam [85] to help solve a new phase of aluminophosphate UiO-7, and in particular, SA was used to predict the number of T sites in the asymmetric part of the unit cell and to identify the framework topology.

4 Conclusions and Perspectives

SA and GA have been very successful in generating desired ionic, closest-packed crystal structures; from the first successful ionic structure to be generated using a SA approach (that of NbF_4 [58]), the structural solution of previously unknown structures for Li_3RuO_4 [2] and $LiCoF_4$ [71], to the prediction of candidate structures which have yet to be synthesised [4, 32, 74, 75]. Although one set of parameters (for both the algorithm and the cost function employed) can generate a large range of different observed structures [1, 74], in general, the algorithm and defined cost function is continuously being refined for the next system(s) of interest. The definition of the cost function can be quite dynamic in that the initial candidate structures produced may influence the user to change the weighted contributions (e.g. the weight for discrepancies in coordination numbers could be increased when the better candidate structures have coordination numbers that are different from those expected).

There are still many unanswered questions; the most common being which is best, a GA or a SA approach? Although both methods were successfully applied to the generation of TiO_2 polymorphs [1, 48, 66, 68], it would be interesting to see how the success rate would change if supercells of the various polymorphs were used such that the number of ions were the same for each polymorph (and the ratio of the unit cell parameters a:b:c as similar as possible). With an increasing number of ions within the unit cell, candidate structures are more likely to contain regions of different desired structural topology as two polymorphs begin to form simultaneously. It is also more likely that the structure will contain point defects or dislocations [69, 70]. It may well be just a simple case of cation disorder, a particular problem when modelling ternary oxides with cations that have similar ionic radii and identical charges [1]. One solution to the multipolymorph candidates is to use the different structural topologies as the starting point for a new SA run [78]. The solution of the cation disorder is to introduce a new move class operator in which there is a probability that two randomly chosen ions are switched [1].

Microporous structures have also been generated using both a GA [18, 83] and SA [84–90] algorithms as well as systematic approaches [5, 25, 26, 36, 38]. In the GA approach EZs, employing both a direct (nonlinear grid) and an indirect (penalise the cost function) method, were used to help generate the porous structures. In a SA approach the structural positions of the cations were generated before the anions (oxide ions) were added. Again different types of cost functions were used. For example, the cost function was based either on the discrepancies in simulated and expected geometrical properties of the cations and powder diffraction data or on a combination of the discrepancies in simulated and expected geometrical properties of the ions and the lattice energy.

There are still avenues for further development of the GA in terms of its performance. Examples include

1. The use of grey binary numbers (where one mutation would always imply a small ionic displacement, rather than the possibility of an ion moving half the way across the unit cell) for the DNA representation of the unknown coordinates towards the end of a GA run.

2. Allowing the GA parameters to be dynamic. For example, the number of grid points N_G and the probabilities P_t, P_c and P_m could vary either as a function of time (GA cycle) or when the diversity of the population falls too low (especially when the GA is stuck near the wrong local minimum of the cost-function hypersurface).
3. Developing ways of helping the GA to simultaneously explore more local minima on the cost-function hypersurface such that the move class crossover is more effective. For example, have several current populations (which evolve by procreation within each separate population), and allow a small probability of candidate structures being able to migrate between populations.

Both GA and SA approaches have recently moved towards using polyatomic units (secondary units, where individual atoms/ions are referred to as primary units) to generate crystal structures [92, 93]. For example, in the multistage procedure used by Woodley et al., the structure of the rigid polyatomic units, described using the Z matrix, forms part of the initial input. Note that polyatomic units like CO_3^{2-}, NH_4^+ and SO_4^{2-} can be detected from the vibrations seen in IR and Raman spectroscopic data and are therefore assumed to be known. In the first stage, where approximate candidate structures are found, a GA is employed to find the position and orientation of the polyatomic units as well as the other ionic coordinates. In the second stage, when the lattice energy is directly minimised (quench stage), the polyatomic units are allowed to "decompose" (treated as separate atoms/ions rather than as a rigid unit). The obvious advantage of using "known" polyatomic units is the reduction in the number of unknown variables. Of course polyatomic units (Z matrices) are already readily used in the crystallographic community where the crystal structures of organic molecular solids are predicted. Definitions of their move class operators and how they sample the orientation of the secondary units is directly applicable to inorganic solids. There are many variations of a polyatomic approach in inorganic structure prediction, like that recently used by Mellot-Draznieks et al. [93], where, for example, a CO_4^{4-} building unit is used in the system Li_4CO_4. Octahedral units, such as MnO_6, and tetrahedral units, such as CuO_4, are commonly seen in inorganic structures. These could be used as polyatomic units provided that somehow these units could be overlapped such that there were the correct number of anions; see for example [94, 95] where "sticky atoms" at short separations are reduced to single atoms. Alternatively, some of the cations could be treated as primary units. For example, when modelling SiO_2, the unit cell would contain the same number of SiO_4 tetrahedra and Si units.

Finally, consider a system whereby some of the ionic coordinates are known. By fixing these ions, different types of applications for the evolutionary algorithms described here are also readily accessible, which include

1. If some of the constituent ions are weak scatterers of X-rays, their distribution may be unknown (e.g. a GA could be used to find the Li^+ distribution in quaternary spinels [96].
2. Fixing the framework of a microporous or a mesoporous material, we can seek the positions of the impurity or adsorbed molecules [97].
3. A GA could be applied to finding the probable ordering of cations within a solid solution, like $La_{1-\delta}Ca_{\delta}MnO_3$ [98].

Acknowledgements Close collaboration with Richard Catlow, Julian Gale and Peter Battle over the past 6 years was vital for our developments in the field of GAs and the new interatomic potentials for modelling open-shell transition-metal ions. This and financial support from the E.P.S.R.C. are gratefully acknowledged. I am also grateful for useful discussions with Alexey Sokol and provision of visualisation software by Accelrys Inc.

5
References

1. Woodley SM, Battle PD, Gale JD, Catlow CRA (1999) Phys Chem Chem Phys 1:2535
2. Bush TS, Catlow CRA, Battle PD (1995) J Mater Chem 5:1269
3. Jansen M, Schön JC (1996) Z Anorg Allg Chem 624:533
4. Wevers MAC, Schön JC, Jansen M (1998) J Solis State Chem 136:233
5. Foster MD (2003) PhD thesis. University of London
6. Alfe D, Gillan MJ, Price GD (2000) Nature 405:172
7. Born M, Huang K (1954) Dynamical theory of crystal lattices. Oxford University Press, Oxford
8. Catlow CRA, Mackrodt WC (1982) (eds) Lecture notes in Physics, vol 166. Springer, Berlin Heidelberg New York
9. Dove MT (1993) Introduction to lattice dynamics. Cambridge University Press, Cambridge
10. Nye JF (2000) Physical properties of crystals, 1985 edn. Oxford University Press, Oxford
11. Gale JD (1996) Philos Mag B73:3
12. Gale JD (1997) J Chem Soc Faraday Trans 93:629
13. Payne MC, Teter MP, Allan DC, Arias AT, Joannopoulos JD (1992) Rev Mod Phys 64:1045
14. Catlow CRA (1997) (ed) Computer modelling in inorganic crystallography. Academic, London
15. Shannon MD, Casci JL, Cox PA, Andrews SJ (1991) Nature 353:417
16. Gale JD, Catlow CRA, Mackrodt WC (1992) Modell Simul Mater Sci Eng 1:73
17. Maddox J (1988) Nature 335:201
18. Woodley SM, Battle PD, Gale JD, Catlow CRA Phys Chem Chem Phys (submitted)
19. Coley DA (1999) An introduction to genetic algorithms for scientists and engineers. World Scientific, Singapore
20. Freeman CM, Gorman AM, Newsam JM (1997) Simulated annealing and structure solution. In: Catlow CRA (ed) Computer modelling in inorganic crystallography. Academic, London
21. Rietveld HM (1969) J Appl Crystallogr 2:65
22. Cheetham AK, Taylor JC (1977) J Solid State Chem 21:22
23. Larson AC, Von Dreele RB (1990) General structure analysis system. Los Alamos National Laboratory Report, LAUR 86
24. Young RA (ed) (1993) The Rietveld method. Oxford University Press, Oxford
25. Price GD, Wood IE, Akporiaye DE (1992) The Prediction of zeolite structures. In: Catlow CRA (ed) Modelling the structure and reactivity in zeolites. Academic, London
26. Klinowski J (1998) Curr Opin Solid State Matter Sci 3:79
27. Friedrichs OD, Dress AWM, Huson DH, Klinowski J, Mackay AL (1999) Nature 400:644
28. Jacobs PWM, Rycerz ZA (1997) Molecular dynamics methods. In: Catlow CRA (ed) Computer modelling in inorganic crystallography. Academic, London
29. Spears WM, De Jong KA (1991) An analysis of multi-point crossover. In: Rawlins G (ed) Foundations of genetic algorithms. Kaufmann, San Francisco
30. Holland JH (1975) Adaptation in natural and artificial systems. University of Michigan Press, Ann Arbor
31. Gottschalk W (1989) Allgemeine Genetik. Thieme, Stuttgart
32. Putz H, Schön JC, Jansen M (1998) Comput Mater Sci 11:309
33. Vainshtein BK, Fridkin VM, Indenbom VL (2000) Modern crystallography: structure of crystals, vol 2, 3rd edn. Springer, Berlin Heidelberg New York

34. Giacovazzo C, Monaco HL (1992) Fundamentals of crystallography. Oxford University Press, Oxford
35. Loewenstein W (1954) Am Mineral 39:92
36. Treacy MMJ, Randall KH, Rao S, Perry JA, Chadi DJ (1997) Z Kristallogr 212:768
37. Baerlocher C, Hepp A, Meier WM (1997) DLS-76 a program for the simulation of crystal structures by geometric refinement. Institut für Kristallographie und Petographie, ETH, Zurich
38. Foster MD, Bell RG, Klinowski J (2001) Stud Surf Sci Catal 135:16P13
39. Harris KDM, Johnston RL, Kariuki BM (1998) Acta Crystallogr Sect A 54:632
40. Schön JC, Jansen M (2001) Z Kristallogr 216:307
41. Schön JC, Jansen M (2001) Z Kristallogr 216:361
42. Lanning OJ, Habershon S, Harris KDM, Johnston RL, Kariuki BM, Tedesco E, Turner GW (2000) Chem Phys Lett 317:296
43. Putz H, Schön JC, Jansen M (1999) J Appl Crystallogr 32:864
44. Catlow CRA (1989) J Chem Soc Faraday Trans 2 85:335
45. Woodley SM, Gale JD, Battle PD, Catlow CRA (2003) J Chem Phys 119:9737
46. Bourova E, Parker SC, Richet P (2000) Phys Rev B 62:12052
47. Woodley SM, Battle PD, Gale JD, Catlow CRA (2003) Chem Mater 15:1669
48. Freeman CM, Catlow CRA (1992) J Chem Soc Chem Commun 89
49. Tosi MP (1964) Solid State Phys 16:1
50. Jackson RA, Catlow CRA (1988) Mol Simul 1:207
51. Woodley SM (2003) Database of published interatomic parameters. http://www.ri.ac.uk/Potentials
52. Schön JC, Jansen M (1994) Ber Bunsenges Phys Chem 98:1541
53. Woodley SM, Battle PD, Catlow CRA, Gale JD (2001) J Phys Chem B 105:6824
54. Dick BG, Overhauser AW (1958) Phys Rev 112:90
55. Woodley SM (2002) XIX congress and general assembly of the international union of crystallography
56. Sanders MJ, Leslie MJ, Catlow CRA (1984) J Chem Soc Chem Commun 1271
57. Jackson RA, Parker SC, Tschaufeser (1992) Lattice energy and free energy minimization of zeolites. In: Catlow CRA (ed) Modelling the structure and reactivity in zeolites. Academic, London
58. Pannetier J, Bassas-Alsina J, Rodriguez-Carvajal J, Caignaert V (1990) Nature 346:343
59. Brown ID (1997) Bond valence methods. In: Catlow CRA (ed) Computer modelling in inorganic crystallography. Academic, London
60. Brown ID (1988) Acta Crystallogr Sect B 44:545
61. Brown ID (1992) Acta Crystallogr Sect B 48:553
62. Wales DJ, Doye JPK (1997) J Phys Chem 101:5111
63. Turner GW, Tedesco E, Harris KDM, Johnston RL, Kariuki BM (2000) Chem Phys Lett 321:183
64. Press WH, Teukolsky SA, Vetterling WT, Flannery BP (1992) Numerical recipes, 2nd edn. Cambridge University Press, Cambridge
65. Bush TS, Battle PD, Catlow CRA, General purpose structure prediction progam (in preparation)
66. Freeman CM, Newsam JM, Levine SM, Catlow CRA (1993) J Mater Chem 3:531
67. Hyde BG, Andersson S (1989) Inorganic crystal structures. Wiley, New York
68. Reinaudi L, Carbonio RE, Leiva EPM (1998) J Chem Soc Chem Commun 255
69. Parker SC, Cormack AN, Catlow CRA (1984) Acta Crystallogr 40:200
70. Reinaudi L, Leiva EPM, Carbonia RE (2000) J Chem Soc Dalton Trans 23:4258
71. Lacorre P, Pannetier J, Hoppe R, Averdunk F, Ferey G (1989) J Solid State Chem 79:1
72. Kirkpatrick S, Gelatt CD, Vecchi MP (1983) Science 220:671
73. Putz H, Schön JC, Jansen M (1995) Ber Bunsenges Phys Chem 99:1148
74. Schön JC, Jansen M (1995) Comput Mater Sci 4:43
75. Schön JC, Jansen M (1996) Angew Chem Int Ed Engl 35:1287
76. Schön JC, Putz H, Jansen M (1996) J Phys Condens Matter 8:143

77. Schön JC (1996) Ber Bunsenges Phys Chem 100:1388
78. Hannemann A, Hundt R, Schön JC, Jansen M (1998) J Appl Crystallogr 31:922
79. Pisani C, Dovest R, Rortti C (1988) Hartree-Fock ab initio treatment of crystalline systems. Springer, Berlin Heidelberg New York
80. Schön JC, Wevers MAC, Jansen M (2001) J Mater Chem 11:69
81. De Leeuw SW, Perram JW, Smith ER (1980) Proc R Soc London Ser A 373:27
82. Wright PA, Natarajan S, Thomas JM, Bell RG, Gai-Boyes PL, Jones RH, Chen J (1992) Angew Chem Int Ed Engl 31:1472
83. Woodley SM, Gale JD, Battle PD, Catlow CRA J Chem Soc Chem Commun (accepted)
84. Deem MW, Newsam J (1989) Nature 342:260
85. Deem MW, Newsam J (1992) J Am Chem Soc 114:7189
86. Falcioni M, Deem MW (1999) J Chem Phys 110:1754
87. Akporiaye DE, Fjellvåg H, Halvorsen EN, Hustveit J, Karlsson A, Lillerud KP (1996) J Phys Chem 100:16641
88. Boisen MB, Gibbs GV, Bukowinski MST (1994) Phys Chem Miner 21:269
89. Teter DM, Gibbs GV, Boisen MB, Allan DC, Teter MP (1995) Phys Rev B 52:8064
90. Boisen MB, Gibbs GV, O'Keeffe M, Bartelmehs KL (1999) Microporous Mesoporous Mater 29:219
91. Meier WM, Villiger HZ (1969) Z Kristallogr 129:411
92. Woodley SM (2000) Energy landscapes. Workshop, Technische Universität Chemnitz
93. Mellot-Draznieks C, Girard S, Férey G, Schön JC, Cancarevic Z, Jansen M (2002) Chem Eur J 8:4102
94. Mellot-Draznieks C, Newsam JM, Gorman AM, Freeman CM, Férey G (2000) Angew Chem Int Ed 39:2270
95. Férey G, Mellot-Draznieks C, Loiseau T (2003) Solid State Sci 5:79
96. Woodley SM, Catlow CRA, Piszora P, Stempin K, Wolska E (2000) J Solid State Chem 153:310
97. Calmiano MD (2001) PhD thesis. University of London
98. Akhtar MJ, Catlow CRA, Slater B, Woodley SM J Mater Chem (submitted)

Structure and Bonding, Vol. 110 (2004): 133–152
DOI 10.1007/b13935

Applications of Evolutionary Computation in Drug Design

Valerie J. Gillet

Department of Information Studies, University of Sheffield, Western Bank, Sheffield S10 2TN, UK
E-mail: v.gillet@sheffield.ac.uk

Abstract Evolutionary algorithms have been widely adopted for solving many of the complex tasks involved in drug design. Here, applications to two different tasks are reviewed: combinatorial library design and deriving quantitative structure-activity relationship (QSAR) models. Combinatorial libraries are the result of combinatorial synthesis whereby robotics is used to synthesise large numbers of compounds in parallel. Many more compounds could be made than can be handled in practice; thus, methods are required for selecting appropriate subsets of compounds. The large numbers involved make this problem well suited to the application of optimisation techniques such as evolutionary algorithms. QSARs attempt to relate a numerical description of molecular structure to known biological activity. Large numbers of easily computable molecular descriptors are now available that could be used to derive models; however, when more descriptors are available than observations, or compounds, overfitting of the data can result where the models generated have poor predictive ability. Thus evolutionary algorithms have successfully been applied to select descriptors that give rise to good QSAR models.

Keywords Combinatorial library design · Quantitative structure – activity relationships · Evolutionary algorithms · Genetic algorithms · Genetic programming

1 Introduction 134

2 Combinatorial Library Design 135

2.1 Library Design Strategies 136
2.2 Molecule-Based Methods 138
2.3 Library-Based Methods 139
2.4 Multiobjective Library Design 141

3 Quantitative Structure-Activity Relationships 143

3.1 Feature Selection Using Evolutionary Algorithms 145
3.2 Feature Selection Using Multiobjective Genetic Programming 146

4 Conclusions 150

5 References 151

Abbreviations

EA	Evolutionary algorithm
GA	Genetic algorithm
GFA	Genetic Function Approximation

GP Genetic programming
MOGA Multiobjective genetic algorithm
MOGP Multiobjective genetic programming
PCA Principal component analysis
QSAR Quantitative structure-activity relationships

1 Introduction

The design of novel compounds that can act as drugs is a complex and expensive process. Most drugs achieve their activity by interfering with the metabolic processes that take place within the body. The disruption to metabolism is often caused by the drug interacting with a biological macromolecule such as an enzyme or receptor and thus preventing the natural substrate from binding. Preferential binding of a drug to a macromolecule requires that there is a high degree of complementarity between the drug and the receptor site: the drug must have complementary shape and electrostatic properties to enable the formation of interactions between the drug and the receptor, such as hydrogen and ionic bonds and lipophilic contacts. As well as having good binding properties, it is also important that a drug is able to reach its site of action within the body, for example, it should be able to pass through one or more physiological barrier, such as a cell membrane, it should remain within the body for an appropriate period of time and it should not be toxic. These effects are often referred to as ADMET (absorption, distribution, metabolism, excretion and toxicity) properties. These properties should be taken into account as early in the drug discovery process as possible.

A wide variety of computational methods have been developed to assist in drug discovery programmes. When the 3D structure of the enzyme or receptor is known then the direct methods of structure-based drug design can be used. Structure-based drug design methods, such as docking [1] and de novo design [2], attempt to identify novel compounds within the context of the binding site of the macromolecule. In most cases, however, the 3D structure of the target is not known and hence indirect methods are required. Indirect methods generally attempt to discover new compounds by using knowledge of known actives. When several actives and inactives are known then an attempt can be made to derive a model of activity, for example, using quantitative structure-activity relationship (QSAR) methods [3] or machine-learning techniques [4]. If several actives are known then it may be possible to identify the 3D arrangement of functional groups required for binding, a process known as pharmacophore identification [5]. The pharmacophore can then be used to search a database to identify other compounds that can adopt the same pharmacophore. When just a single active is known, similarity methods can be used to identify structurally similar molecules in the hope that they will have similar activities [6].

The traditional approach to drug discovery involves an iterative synthesis and test cycle in which a medicinal chemist synthesises a compound, tests it for activity and then uses the results to suggest a new compound for synthesis, in the next iteration. However, a revolutionary change has taken place in the drug-discovery process over the last decade whereby the use of automation techniques has become commonplace [7]. Thus, robotics is now used routinely both to screen compounds for biological activity, in a process known as high-throughput screening, and also to synthesise large numbers of compounds simultaneously, in a process known as combinatorial synthesis. Combinatorial synthesis allows tens or even hundreds of thousands of compounds to be made in a single experiment, in what are known as combinatorial libraries, and it is now possible to test similar numbers of compounds in a single high-throughput screening experiment. Thus, the throughput of the synthesis and test cycle has increased enormously. Despite this increased throughput, the number of compounds that could potentially be made is much larger than could ever be handled practically (it has been estimated that as many as 10^{40} "druglike" compounds could exist [8]) and hence to be effective combinatorial libraries and high-throughput screening experiments must be designed very carefully.

The explosion of data available about compounds and their activities together with increases in computing performance have resulted in the development of virtual screening techniques. Virtual screening is the in silico equivalent of high-throughput screening and involves the computational analysis of databases of compounds, either real or virtual (e.g., compounds that could be synthesised using combinatorial synthesis), to identify possible candidate molecules [9, 10]. Here, the term virtual screening includes structure-based approaches that require the 3D structure of a binding site as well as approaches such as similarity methods that do not rely on binding site knowledge.

The complex nature of many of the processes involved in drug discovery makes them well suited to the application of evolutionary algorithms (EAs). Since the early 1990s many different applications in computer-aided drug design have been developed, mostly employing genetic algorithms (GAs). For example, GAs have been applied to docking, de novo design, pharmacophore identification, similarity searching, the derivation of QSAR models and combinatorial library design [11]. This chapter focuses on the applications of EAs to two areas of drug design: the design of combinatorial libraries and the identification of QSARs.

2
Combinatorial Library Design

Combinatorial synthesis is the process by which multiple compounds are synthesised simultaneously. Consider a two-component reaction such as the coupling of amines and carboxylic acids to form amides, as shown in Fig. 1. If there are 100 different amines and 100 different carboxylic acids available then combinatorial synthesis offers the potential to synthesis all 100×100 (10^4) products in a single experiment. In contrast, using traditional synthesis in which one compound is made at a time, a medicinal chemist will typically synthesise 50 compounds a year. Thus it is clear that combinatorial methods can result in much greater numbers of compounds available for testing than can be provided using traditional methods.

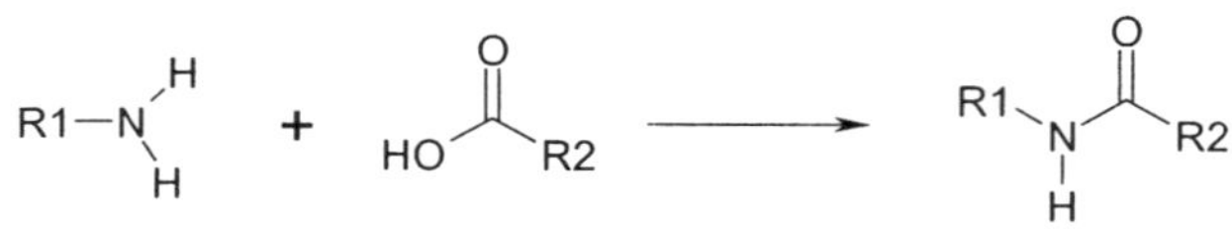

Fig. 1 A combinatorial library of amides can be constructed from amines and carboxylic acids

In most combinatorial synthesis experiments there are many more reactants available than can actually be handled in practice. For example, many thousands of amines and carboxylic acids exist as potential reactants. Thus, rather than attempting to synthesis all possible compounds, methods are required for the selection of reactants so that the resultant library is of a manageable size.

The initial emphasis in combinatorial chemistry was on the synthesis of as many diverse compounds as possible, in the expectation that a large number of molecules would inevitably lead to more hits in the biological assays. However, early combinatorial libraries gave disappointing results with lower hit rates than expected and often the hits that were found had properties that made them unsuitable as lead compounds, for example, they were too large, too insoluble, or contained inappropriate functional groups. Thus, it is now realised that the compounds within combinatorial libraries should have druglike physicochemical properties so that they can be progressed to later stages of the drug design process. Furthermore, the current emphasis is on the design of smaller, more focussed libraries that incorporate as much information about the therapeutic target as possible. For example, if the 3D structure of the target protein is available, then docking methods can be used to try to ensure that the compounds synthesised could potentially bind to the active site. If some active compounds are known then libraries may be designed using pharmacophoric approaches, or using similarity methods that are based on 2D descriptors or physicochemical properties. When little is known about the target then more diverse libraries may be relevant. Such libraries are designed to give a broad coverage of chemistry space so that they can be screened against a range of structural targets.

2.1 Library Design Strategies

The two main strategies for combinatorial library design are known as reactant-based design (also known as monomer-based design) and product-based design [12]. In reactant-based design, reactants are chosen without taking into account the product molecules that will be produced. Thus, in the design of diverse libraries, diverse subsets of reactants are chosen in the hope that they will result in a diverse set of products. Examining all possible subsets of a given size is usually computationally infeasible: there are

$$\frac{N!}{n!(N-n)!} \tag{1}$$

subsets of size n contained within a larger set of N compounds, for example, there are more than 10^{13} subsets of size 10 contained within 100 compounds. Thus a number of methods have been developed that can be used to find approximate solutions. These compound-selection methods include clustering, partitioning and dissimilarity-based compound selection [13–15].

Compound-selection methods are based on a definition of molecular similarity or dissimilarity. This, in turn, requires first that the molecules are characterised using molecular descriptors and second the use of a similarity coefficient to quantify the degree of similarity based on the descriptors. Many different types of descriptor have been suggested for calculating structural similarity and measuring diversity [16–18]. The most commonly used descriptors are whole molecule properties, such as molecular weight and lipophilicity, descriptors derived from the 2D representation of molecules, such as topological indices and fragment-based fingerprints, and descriptors that represents molecules in 3D, such as pharmacophore keys. Whole molecule properties and topological indices are usually represented as real-numbered vectors. Fragment-based 2D fingerprints and pharmacophore keys record the presence or absence of fragment substructures or pharmacophoric patterns, respectively, within a molecule as a binary fingerprint. When molecules are represented by binary fingerprints, similarity is usually quantified using an association coefficient such as the Tanimoto coefficient, whereas when molecules are described by physicochemical properties or other real-valued descriptors, then similarity is usually quantified using a distance coefficient, typically, Euclidean distance [6]. The similarities or dissimilarities between molecules provide the input to the various subset selection methods.

In product-based design, the properties of the resulting product molecules are taken into account when selecting the reactants. The first step is usually to fully enumerate the virtual product library. Product-based design is thus much more computationally demanding than reactant-based design: if there are 100 reactants available for each reactant pool in a two-component reaction then there are 10^4 products in the fully enumerated library and there are more than 10^{240} subsets of size 100 contained within a library of this size. Thus, product-based selection is well suited to the application of EAs and several different methods have been developed [19, 20], examples of which are given later. Product-based approaches to library design can be divided into molecule-based methods and library-based methods [21]. In molecule-based methods, individual product molecules that have the desired properties are selected in a first step and combinatorial libraries are designed in a second step by identifying reactants that occur frequently within the desired products. Other approaches are library-based and involve selecting a combinatorial subset directly such that every reactant selected for one component occurs in a product with every reactant selected from the other components, i.e., the combinatorial constraint is taken into account. The number of combinatorial subsets in a virtual library is

$$\prod_{i=1}^{R} \frac{N_i!}{n_i!(N_i - n_i)!}, \qquad (2)$$

where there are R components or sets of reactants and there are n_i reactants to be selected from a possible N_i for each component. Thus, there are approximately

10^{26} different 10×10 libraries that could be synthesised from a 100×100 virtual library.

Despite the greater computational complexity of product-based methods relative to reactant-based methods they can be more effective, especially when the aim is to optimise the properties of a library as a whole, such as diversity or the distribution of physicochemical properties [12,22,23]. In addition, product-based selection is usually more appropriate for focussed libraries which require consideration of the properties of the product molecules themselves.

2.2 Molecule-Based Methods

One of the earliest approaches to product-based design that is based on a GA is the method described by Sheridan and coworkers [21,24]. Each chromosome of the GA encodes a product molecule as the list of reactants from which it is made. Thus, crossover and mutation involve changing some of the reactants to give rise to a new product molecule. Molecules can be optimised via a variety of different fitness functions, including similarity to a target molecule using atom-pair descriptors (a variation on 2D fragment-based fingerprints), and fit to a receptor site, which involves generating 3D conformations and docking them within the receptor site. When the GA has terminated the entire population of chromosomes is analysed to identify reactants that occur frequently across all the molecules in the population. The frequently occurring reactants could then be used to construct a combinatorial library. The original publication describes the design of a tripeptoid library with three positions of variability: 2,507 amines were available for two of the substitution positions and 3,312 for the third position. This gives a virtual library of approximately 20 billion potential tripeptoids. The method was able to find molecules that are very similar to given target molecules after exploring only a very small fraction of the total search space.

The fitness functions embodied in molecule-based methods are relatively fast, especially when optimisation is based on 2D properties, since they typically involve a pairwise molecular comparison with a target molecule rather than the analysis of an entire library, as is the case in library-based methods. However, there is no guarantee that building libraries from frequently occurring reactants will result in optimised libraries, nor is it possible to optimise properties of the library as a whole.

A similar approach has been developed in the program Focus-2D [25, 26], where molecules are described using MolconnX topological descriptors and they are evolved to be similar to a known target compound or to have high predicted activity based on a precomputed QSAR. Both a GA and simulated annealing have been implemented as optimisation techniques. In this method, the second step of identifiying reactants (or monomers) that occur frequently in the selected product molecules has been termed Monomer Frequency Analysis.

Weber et al. [27] developed a strategy for library design that is based on a GA and uses a fitness function based on experimentally determined activities. Thus, the fitness function requires the actual synthesis and biological testing of compounds. The approach was tested on the Ugi reaction, which is a four-component reaction. The virtual library consisted of 160,000 possible products that could be made from 10 isonitriles, 40 aldehydes, 10 amines and 40 carboxylic acids. The approach is molecule-based and uses a chromosome representation that is similar to that used by Sheridan and coworkers with the individual reactants that make up a product molecule encoded in the chromosome. The GA was initialised with a population of 20 randomly assigned chromosomes. Each chromosome was then scored by synthesising and testing the compound it represented. The best product molecule in the initial population exhibited an IC_{50} of 300 µM. A thrombin inhibitor with submicomolar IC_{50} was found after just 20 generations of the GA, i.e., after synthesising and testing just 400 molecules. In a follow-up study [20] a full combinatorial library of 15,360 products was synthesised from a three-component Ugi reaction scheme using 12×8×60 substituents and all the products were tested for activity against serine protease thrombin. The resulting structure-activity data were then used to investigate the behaviour of various GAs parameters, including different encoding schemes, mutation versus crossover rates and population sizes. Similar approaches also based on experimentally determined fitness functions were published at about the same time by Singh et al. [28] and Yokobayashi et al. [29].

Gobbi and Poppinger [30] developed a molecule-based method that uses a different chromosome representation. In their GA, each chromosome represents a molecule as a binary fingerprint. As already mentioned, a fingerprint is a binary string in which each bit represents a particular substructure fragment. For a given molecule a bit is set to 1 if the fragment it represents is contained in the molecule, otherwise it is set to 0. The advantage of this approach compared to the earlier approaches is that the children produced by applying crossover and mutation do not have to contain the same reactants as their parents and also they are not limited to the same reaction. However, crossover and mutation can generate molecules that do not exist in the virtual library and chromosomes can be produced that are chemical nonsense since the fragments or bits in the binary string are not independent. The fitness function involves finding the molecule in the library whose fingerprint is most similar to the chromosome and then testing the molecule for activity. Rather than enumerating the entire virtual library and calculating descriptors for all the compounds, which may not be possible for very large libraries (over 1 million compounds), the fitness function samples the virtual library using a TABU search method to find a compound that is similar to the chromosome. The method was tested in simulated screening experiments that involved collections of molecules where the activities were already known and was able to find all the active compounds by screening approximately 10% of the datasets, this representing a 100-fold improvement on random selection.

2.3 Library-Based Methods

Library-based methods attempt to optimise a combinatorial subset directly. Although the computational complexity of the subset selection problem is reduced owing to the combinatorial constraint, the fitness functions employed in library-based methods are more demanding than those used in molecule-based methods since they require the analysis of an entire subset of molecules rather than just a single compound.

Brown and Martin [31] described a library-based GA for combinatorial library design in a program called GALOPED. Their method was developed for the design of diverse combinatorial libraries synthesised as mixtures. The mixtures approach to combinatorial synthesis uses a technique known as split-and-mix, where several compounds are synthesised and screened in the same vessel. (Parallel synthesis, on the other hand, involves the synthesis of compounds as discretes, where there is one compound per vessel.) The synthesis of mixtures allows much higher throughputs to be achieved than parallel synthesis; however, if activity is seen in a vessel the mixture must be deconvoluted with the individual compounds contained within it synthesised and then tested to identify the particular compound that is responsible for the activity. Deconvolution can be achieved using mass spectroscopy techniques where the amount of resynthesis and testing is minimised by reducing the redundancy in molecular weights of the compounds contained in a single vessel. GALOPED attempts to optimise mixtures based on their diversity and ease of deconvolution.

Each chromosome encodes a combinatorial subset as a binary string. The chromosome is partitioned with one partition for each component, or substitution position, in the library. The number of bits in the chromosome is equal to the sum of reactants available in each reactant pool so that each bit represents a different reactant. Thus a virtual library of 1,000×1,000 potential products will require chromosomes with 2,000 bits. A bit value of 1 indicates that a reactant is included in the combinatorial subset and a value of 0 indicates that the reactant has not been selected. The size of the subset selected can vary according to the number of bits set to 1 and so minimum and maximum thresholds are set by the user and libraries outside the desired size range are penalised in the fitness function. The fitness function involves maximising the diversity of the library while minimising the molecular weight redundancy. In GALOPED, diversity is measured by first enumerating the library represented by a chromosome and then clustering it based on 2D descriptors and counting the number of different clusters occupied by the library. Clustering is a computationally expensive process and thus the size of combinatorial libraries that can be handled by this method is limited.

Other approaches are based on designing a fixed-sized combinatorial library. For example, the SELECT program [32] is based on a GA in which each chromosome encodes one possible fixed-sized combinatorial subset contained within the virtual library. Thus, for a two-component combinatorial synthesis in which n_A of a possible N_A first reactants are to be combined with n_B of a possible N_B second reactants, the chromosome of the GA contains n_A+n_B integers (cf. the N_A+N_B binary encoding scheme used in GALOPED). Each of the n_A integers represents a reactant selected from the N_A that are available in pool A and each of the n_B integers represents a reactant selected from the N_B that are available in pool B. Thus, in SELECT the size of the combinatorial subset to be selected is specified on input. The fitness function quantifies the "goodness" of the combinatorial subset encoded in the chromosome and the GA evolves new potential subsets in an attempt to optimise this quantity. A variety of different fitness functions are available allowing the design of libraries that are diverse or focussed, for example, the fitness function can be based on calculating pairwise molecular similarities or dissimilarities or on maximising or minimising the coverage of the library in some partitioned chemistry space. Libraries can also be optimised on multiple properties via a weighted-sum fitness function as described in the next section.

Lewis et al. [33] have developed both simulated annealing and GA approaches to selecting combinatorial subsets from virtual libraries. The GA version was used to design a subset of a benzodiazepine library of configuration 4×4×3×2 selected from a virtual library of 11×7×5×4 products. The GA was designed to maximise the coverage of pharmacophores in the subset compared with the coverage of the full virtual library. Generation of the pharmacophore descriptors is computationally expensive since it involves a full conformational analysis of the virtual library; hence, the size of libraries that can be handled by this method is limited compared with methods that are based on 2D descriptors.

2.4 Multiobjective Library Design

The approaches to library design described so far have generally been concerned with the optimisation of a single objective, such as diversity or similarity to a known active compound. However, it is usually desirable to optimise multiple properties simultaneously. For example, in addition to designing a library that is diverse or focussed it is usually desirable that the compounds contained within the library have druglike physicochemical properties and can be made from readily available, inexpensive reactants, and so on. Such a situation is referred to as multiobjective library design.

Most approaches to tackling this problem make use of a weighted-sum fitness function, such as the one in Eq. (3) that is used in the SELECT program [32].

$$f(n) = w_1 \cdot \text{diversity} + w_2 \cdot \text{cost} + w_3 \cdot \text{property1} + w_4 \cdot \text{property2} + \ldots \quad (3)$$

Typically, SELECT would be configured to design libraries that simultaneously have maximum diversity, minimum cost and druglike physicochemical properties. The physicochemical property profiles are optimised by minimising the difference between the distribution of a property in the library and some reference distribution, for example, the distribution of the property in a collection of known drugs. Each of the properties is standardised and then relative weights are defined by the user at run time.

Several other library design programs also use a weighted-sum fitness function for the simultaneous optimisation of multiple properties [33–36]; however, there are limitations associated with this approach especially when the objectives are in conflict, which is usually the case. For example, it can be difficult to determine appropriate weights for the objectives and typically several trial-and-error runs are required where the relative weights are varied. Each set of weights will result in a different solution that represents a different compromise in the ob-

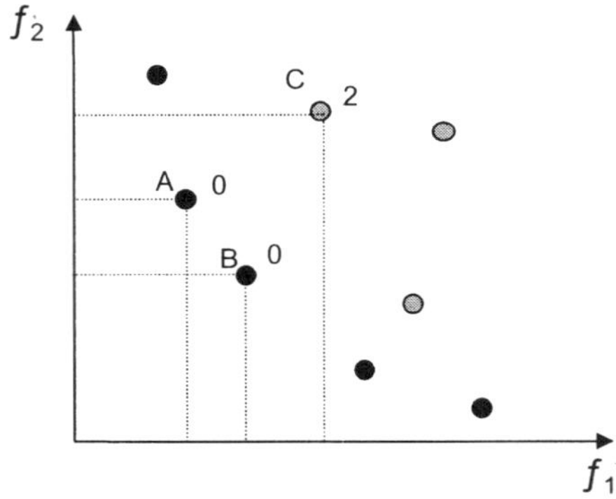

Fig. 2 Pareto ranking and dominance for a two-objective problem, minimising f_1 and f_2. A nondominated individual is one for which there is no other individual that is better in all objectives. In Pareto ranking, an individual's rank corresponds to the number of individuals in the current population by which it is dominated. Solutions A and B are given rank 0 since they are nondominated, whereas solution C is given rank 2 since it is dominated by both A and B

jectives. Thus, in fact a family of solutions usually exists where, in the absence of further information, all solutions can be seen as equivalent but where they each represent a different compromise in the objectives.

EAs, including GAs, are well suited to multiobjective optimisation since they operate on a population of individuals and hence they can be easily adapted to search for multiple solutions in parallel. Fonseca and Fleming [37–39] have developed an approach to multiobjective optimisation known as multiobjective GA (MOGA). The method treats each objective independently without summation and without the need to choose relative weights. In MOGA, a set of nondominated solutions (also known as Pareto-optimal solutions) is sought rather than a single solution. A nondominated solution is one for which no other solution exists in the population that is better in all objectives and one solution dominates another if it is either equivalent or better in all the objectives and, strictly, it is better in at least one objective. The weighted-sum fitness function used in a traditional GA is replaced by fitness based on dominance and each individual is given a rank according to the number of individuals in the population by which it is dominated, in a process known as Pareto ranking, illustrated in Fig. 2. The fitness of an individual is then based on rank with all individuals that have the same rank having the same fitness value and individuals with the lowest ranks (least dominated) being preferred. The result is a family of solutions, with each solution representing a different compromise in the individual objectives. These solutions define the Pareto surface.

The MOGA approach has been adopted in the library design program MoSELECT [40–42]. The result of applying MoSELECT to the design of amide libraries is shown in Fig. 3, where the libraries are optimised on diversity and molecular weight profile. The aim was to maximise diversity (measured as the sum of pairwise dissimilarities calculated using 2D fingerprints and the cosine coefficient [6]) while minimising Δ MW (which is the difference in the distribu-

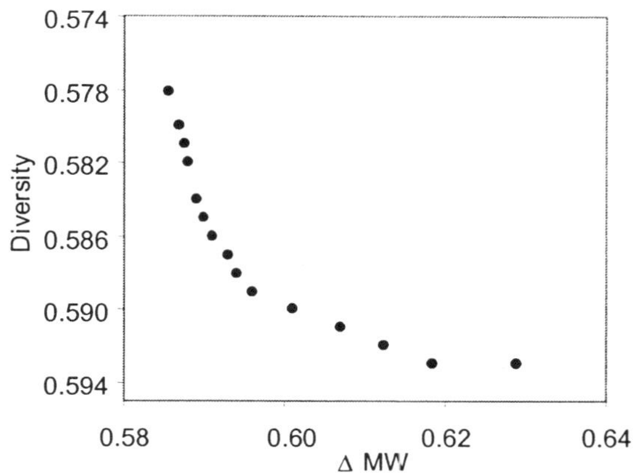

Fig. 3 A family of nondominated solutions is generated that is evenly spread throughout the objective space. ΔMW is the root-mean standard deviation between the molecular weight profile of the library and the profile found in a collection of known drugs

tion of molecular weights in the library compared to a reference distribution calculated for a collection of known drugs). It can be seen that the two objectives are in competition, with more diverse libraries having relatively poor molecular weight profiles and vice versa. The set of different compromise solutions has been found in a single run of the program.

The MOGA represents a significant improvement over the use of a traditional GA for the design of combinatorial libraries that are optimised on multiple objectives. Many of the limitations of the weighted-sum approach have been overcome, for example, different objectives are easily handled; there is no need to assign relative weights to the various objectives; and the trade-off in the objectives can be explored in a single run, thereby removing the trial-and-error approach that is required with a weighted-sum fitness function. The result is a family of different compromise solutions from which the user can make an informed choice on which solution(s) to explore rather than proceeding with a single somewhat arbitrary solution generated using a traditional GA.

3 Quantitative Structure-Activity Relationships

QSARs attempt to relate a numerical description of molecular structure to known biological activity. They are typically derived in cases where biological activities are available for a series of molecules but the 3D structure of the target protein is unknown. By establishing a relationship between structure and activity a good QSAR model can be used to enhance understanding of biological mechanisms and to make predictions about compounds with unknown activities. The approach was pioneered by Hansch in the 1960s, who demonstrated that biological activity could be correlated to a few simple thermodynamic or electronic variables using a simple regression equation [43].

The simplest linear regression equation has the form

$$y = mx + c, \tag{4}$$

where y is known as the dependent variable and is the property being modelled, such as biological activity; and x is the independent variable, for example, a molecular descriptor such as log P, which is the logarithm of the partition coefficient between 1-octanol and water. The aim of the linear regression is to find values for the coefficient m and the constant c that minimise the sum of the differences between the predicted values of activity and the observed values. In QSAR, there is usually more than one independent variable or descriptor, in which case the coefficients and constants can be determined using multiple linear regression.

The quality of a multiple linear regression is usually assessed using the squared correlation coefficient, written r^2 [44]. r^2 is calculated using the following formula:

$$r^2 = 1 - \frac{\sum_{i=1}^{N} (y_i - y_{calc,i})^2}{\sum_{i=1}^{N} (y_i - \langle y \rangle)^2}, \tag{5}$$

where there are N data points or compounds, $\langle y \rangle$ is the mean of the dependent variable, $y_{calc,i}$ is the activity of compound i calculated using the model and y_i is its observed activity. r^2 takes a value between 0 and 1, and represents the proportion of the dependent variable that is explained by the model, with an r^2 of 0 indicating that none of the variation in the dependent variable is explained by variation in the independent variables, and a value of 1 corresponding to a perfect explanation. Thus r^2 measures how well the model fits the data it was trained on.

The quality of a model can also be assessed according to its ability to predict the activities of unknown compounds using a technique known as cross-validation. Cross-validation involves systematically removing one or more data points from the training set, constructing a model using the reduced set and then using the model to predict the data that have been removed. The simplest form is leave-one-out cross-validation, in which a single point is removed. Repeating this for every data point leads to a cross-validated r^2, more usually known as q^2, which provides an estimate of the predictive ability of the model. q^2 is calculated using:

$$q^2 = 1 - \frac{\sum_{i=1}^{N} (y_i - y_{pred,i})^2}{\sum_{i=1}^{N} (y_i - \langle y \rangle)^2}, \tag{6}$$

where $y_{pred,i}$ is the value predicted for compound i.

Since the first analysis by Hansch two significant developments have been made. The first is that a wide range of easily computable molecular descriptors are now available and the second is that many sophisticated techniques have emerged that are a significant improvement over the original linear regression analysis [45]. However, despite these developments, deriving models that are accurate, reliable and easily interpretable remains a difficult task. While the availability of large numbers of easily computable descriptors such as topological indices, substructural keys, and 2D and 3D fingerprints [16, 17], can help in providing a variety of different ways of describing structures it can also make the task of deriving accurate and easily interpretable QSAR models harder. When there are more descriptors available than observations, or compounds, overfitting of the data can result where the models generated have poor predictive ability. In such cases, the number of descriptors should be reduced in order to develop a model that is predictive and easier to interpret. This process is often known as feature selection. A number of different algorithms have been developed for feature selection, including principal component analysis (PCA), stepwise approaches and EAs [45].

In PCA the input descriptors are transformed into orthogonal principal components with a small number of principal components usually being sufficient to represent most of the variation in the original data. PCA can be very useful when the descriptors are correlated since by definition the new set of descriptors will be uncorrelated; however, a disadvantage of the approach is that it can be difficult to interpret the resulting models since each new descriptor represents a linear combination of the original descriptors.

Two stepwise approaches to feature selection are forward-stepping and backward-elimination. In forward-stepping, a model is initially generated using a single descriptor, with the best model being the one with the best statistics. The second descriptor is then selected as the one that gives the best two-term model, with each descriptor being tried in turn. Subsequent descriptors are chosen in the same way until no further improvement in the model is achieved. In backward-elimination, the complete set of descriptors is used initially, with each descriptor being removed in turn. These approaches are limited since they do not take account of coupled effects between descriptors and oweing to their stepwise nature they may lead to a suboptimal subset of descriptors, for example, a descriptor that is eliminated early on may turn out to be informative when other descriptors are eliminated. Hence a number of other approaches have been developed.

3.1 Feature Selection Using Evolutionary Algorithms

The optimisation of a subset of features can be achieved using a GA. Thus an appropriate chromosome representation could be a binary string in which each element represents a different feature. Features can then be included or excluded from the model by assigning the corresponding element the value 1 or 0 respectively. Thus, Rogers and Hopfinger [46] developed the Genetic Function Approximation (GFA) method, where descriptor selection is performed using a GA and QSAR models are obtained by performing least-squares regression to regenerate the coefficients [46]. It is well known that r^2 tends to increase as the number of variables increases, so more complex models tend to fit the data better; however, as already discussed, overfitting can result where the model is poor in prediction. Thus in GFA, models are scored using Friedman's lack of fit measure, which is based on the least-squares error combined with a user-definable smoothing parameter that penalises the effect of including additional terms in the model. The method was tested using the Selwood data set [47], which has become a standard test set for QSAR applications and consists of 31 compounds, 53 descriptors and a set of corresponding antifilarial antimycin activities, expressed as $-\log(IC_{50})$ values.

The mutation and selection uncover models (MUSEUM) algorithm [48, 49] developed by Kubinyi is based on an EA involving mutation only (i.e., there is no crossover operator). It avoids the need for a user-defined parameter by using the FIT value as the fitness criterion. The FIT value is based on the Fischer significance value adjusted with respect to the number of independent variables selected in each model. In the related evolutionary programming method developed by Luke [50], fitness is defined using a three-term function. The first term is the root mean square between predicted and measured values, the second term is used to drive the solution towards a given number of descriptors and the final term is used to weight the descriptors according to their exponent values, for example quadratic terms are penalised relative to linear terms.

So and Karplus [51] have developed a hybrid method that combines a GA for descriptor selection with an artificial neural network for model building. They found improved models for the Selwood data set when compared with the GFA and evolutionary programming methods, with the success being attributed to the ability of the neural network to select nonlinear descriptors.

3.2 Feature Selection Using Multiobjective Genetic Programming

The recently developed MoQSAR program treats QSAR as a multiobjective optimisation problem that comprises a number of competing objectives [52]. The aim is to explore the trade-off between factors such as model accuracy, model complexity and model interpretability. A variety of models are generated, ranging from more accurate models, which tend to be more complex and hence harder to interpret, through to simpler models, that are easier to interpret but where this may have been achieved through the loss of some statistical robustness. Multiobjective genetic programming (MOGP) is used to derive a family of equivalent solutions where each solution represents a different compromise in the objectives. The medicinal chemist can then choose an appropriate compromise solution, for example, by trading statistical robustness for chemical interpretability.

MOGP is based on the more traditional optimisation method genetic programming (GP), which is a type of GA [53, 54]. The main difference between GP and a GA is in the chromosome representation: in a GA an individual is usually represented by a fixed-length linear string, whereas in GP individuals are represented by treelike structures; hence, they can vary in shape and size as the population undergoes evolution. The internal nodes of the tree, typically represent mathematical operators, and the terminal nodes, typically represent variables and constant values: thus, the chromosome can represent a mathematical expression as shown in Fig. 4.

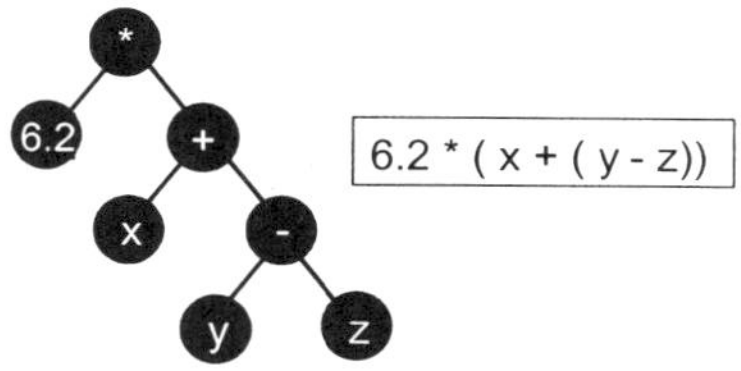

Fig. 4 In genetic programming (*GP*), a chromosome is a tree structure and can be used to represent a mathematical expression where the internal nodes are mathematical operators and the terminal nodes are variable or constant values

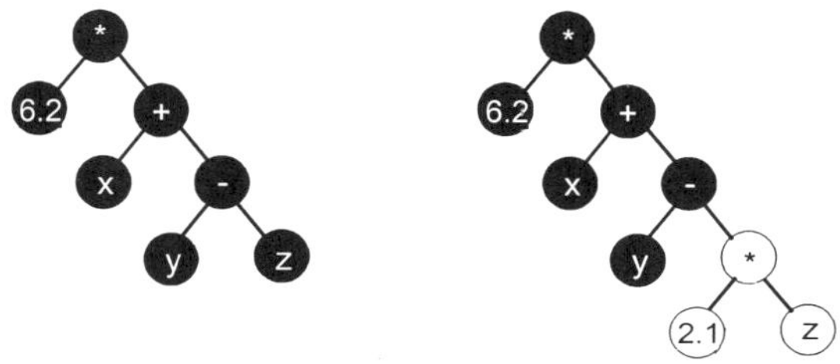

Fig. 5 The mutation operator in GP involves selecting a node at random in a single parent. The subtree at that node is deleted and a new subtree is grown at that point

New chromosomes are evolved by applying the genetic operators reproduction, mutation and crossover. The reproduction operator involves selecting one parent chromosome which is copied unchanged into the next generation. In mutation, a single parent is selected and a mutation point is chosen at random, as illustrated in Fig. 5. The subtree at the mutation point is deleted and a new subtree is grown. In crossover, two parents are selected which are usually of different shape and size, a crossover point is chosen at random in each parent and subtrees are exchanged, as illustrated in Fig. 6. Crossover is the predominant operator

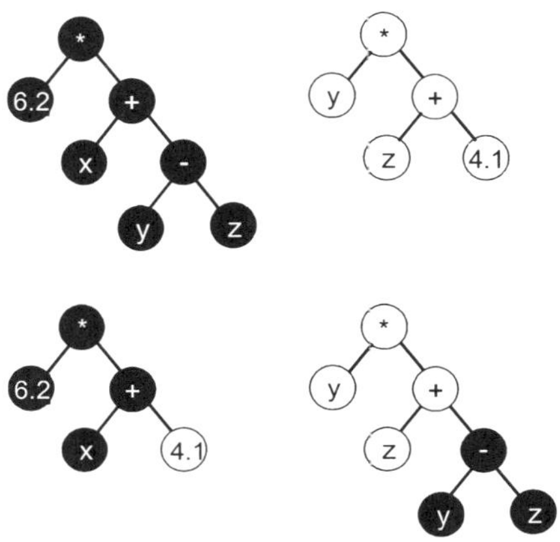

Fig. 6 The crossover operator in GP involves selecting two nodes at random, one in each parent, and exchanging the subtrees headed by each node

in GP and is performed with a high probability relative to mutation and reproduction. The iterations continue until some convergence criterion has been reached when the best solution found is designated to be the result of the GP.

In MoQSAR, the internal nodes include the sum, quadratic and cubic power operators and the terminal nodes consist of the molecular descriptors available for the dataset. A chromosome is translated into a QSAR in two steps: (1) the expression encoded in a chromosome is extracted to determine the descriptors that will be used in the QSAR model; (2) optimum values for the coefficients and the intercept are calculated using the least-squares method.

Traditional optimisation methods, such as GAs, GP and simulated annealing, typically handle multiobjective optimisation problems by reducing multiple objectives to a single objective through the use of a weighted-sum fitness function. The disadvantages of this approach have already been mentioned in the context of combinatorial library design and include the fact that it can be difficult to set appropriate weights for the objectives and only a single solution is found that represents one particular compromise in the objectives.

In MoQSAR, MOGP [52, 54] is used to overcome the limitations of using a weighted-sum fitness function. The approach is similar to the MOGA approach described earlier where multiple objectives are handled independently without summation and without weights and Pareto ranking is used to identify a Pareto-optimal set of solutions. Pareto ranking was shown previously in Fig. 2.

MoQSAR has been demonstrated on a two-objective optimisation problem where the aim is to maximise model fitting, measured by r^2, while simultaneously minimising model complexity, measured by the number of terms, p. When applied to the Selwood dataset with p allowed to vary between 1 and 7, a family of seven independent models ($0.371 \le r^2 \le 0.835$) was found, with one solution for each distinct number of terms. The Pareto surface is shown in Fig. 7, where the trade-off between r^2 and the number of terms is evident, with the more complex models having a better fit to the data. The best four-term QSAR model found includes a cubic term (SURF_A) and a quadratic term (SUM_R) and is shown in Eq. (7). Note that whereas r^2 has been optimised, q^2 was calculated for the model with highest r^2, i.e., it was not included in the optimisation process itself.

$$\begin{aligned} -\log(EC_{50}) = {} & 0.67183\ \mathrm{LOGP} - 2.8519 \times 10^{-8}\ (\mathrm{SURF\_A})^3 \\ & + 1.8824\ \mathrm{SUM\_F} + 17.485\ (\mathrm{SUM\_R})^2 + 3.68537 \end{aligned} \quad (7)$$

$$r^2 = 0.830; \quad q^2 = 0.782$$

In general, the presence of higher-order relationships in QSAR models is undesirable; hence, an additional objective was included in the search, namely the number of nonlinear terms, s. This time, the goal of the optimisation was to maximise r^2 while minimising p and s. Including the third objective, s, results in more than one QSAR model for a given number of terms, with the total number of solutions increasing from seven for the two objective case to 16 for three objectives. The solutions include the best two- to five-term linear models reported in the literature for the Selwood data set [47]. Three four-term models were identified: one consisting of linear terms only (with s=1); one including a quadratic terms

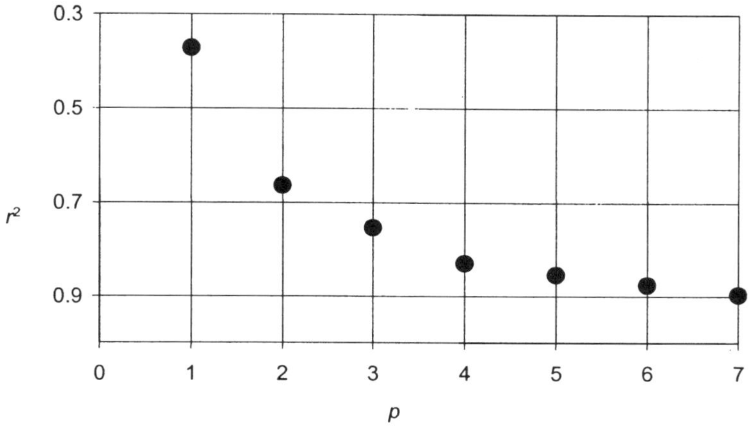

Fig. 7 Solutions found when optimising model fitting (r^2) simultaneously with model complexity, measured by the number of terms in the model, p

(with s=2); and one including a cubic term (with s=3). All of these models represent simpler models than the previously reported four-term model, when optimising r^2 and p only. The linear model is shown in Eq. (8):

$$-\log(\mathrm{EC}_{50}) = 0.49984\ \mathrm{LOGP} - 2.8075\ \mathrm{ATCH4} + 0.84222\ \mathrm{ESDL3} - 0.19960\ \mathrm{PEAK\_X} + 1.7908 \quad (8)$$

$$r^2 = 0.774; \quad q^2 = 0.636$$

The model includes the well-understood descriptor log$P$ and a descriptor based on atomic charge (ATCH4), which is also readily interpretable. However, it also contains the hard-to-interpret descriptors ESDL3 and PEAX_X. Thus, although the model is statistically sound, it does not promise to engage the chemist's imagination.

MoQSAR was subsequently used to investigate the effect of including chemical desirability, D, as an objective. The molecular descriptors were assigned numerical desirability values corresponding to excellent, fair and poor. Examples of descriptors rated as excellent are dipole moment (DIPMOM), molecular weight (MOL_WT), van der Waals volume (VDWVOL), surface area (SURF_A) and LOGP. These are all descriptors that are intuitive to the chemist. When optimising desirability simultaneously with r^2, p and s, a family of 44 QSARs was found, with the number of terms in the models varying from two to eight. The objective values for the four-term linear models are shown in the parallel coordinates plot in Fig. 8, with each line representing a different QSAR model. Each objective has been normalised to be in the range 0–1 and has been plotted such that the direction of improvement is towards 0 on the y-axis. This time a linear four-term model is found in which all the descriptors are rated as excellent (Eq. (9)); however, the model has slightly poorer statistics than the model found previously.

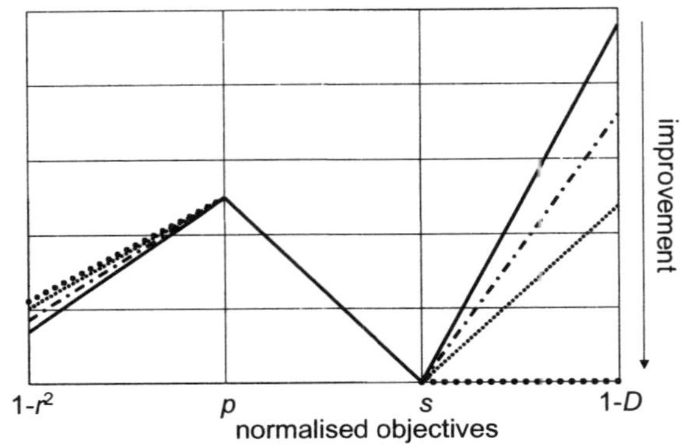

Fig. 8 The objective values for the four-term linear models

Thus, model accuracy has been traded in order to derive a model that is easy to interpret.

$$-\log(EC_{50}) = 0.46825\ \mathrm{LOGP} - 1.9043\times10^{-2}\ \mathrm{VDVWOL} + 7.0068\times10^{-3}\ \mathrm{MOL\_WT} + + 1.3645\ \mathrm{SUM\_F} + 0.14079 \quad (9)$$

$r^2 = 0.730$; $q^2 = 0.616$

MoQSAR represents a new way of deriving QSARs. QSAR is treated as a multiobjective optimisation problem that comprises a number of competing objectives, such as model accuracy, complexity and chemical interpretability. The result is a family of QSAR models where each model represents a different compromise in the objectives. Typically, MoQSAR is able to find models that are at least as good as those found using standard statistical methods. The method will also find models where accuracy is traded with other objectives such as chemical interpretability. When presented with the full range of models the medicinal chemist is able to select one that represents the best compromise over all objectives.

4
Conclusions

Over the last decade EAs have been applied to many of the tasks involved in computer-aided drug design [11]. This chapter has aimed to review the application of EAs in two rapidly developing areas: namely, the design of combinatorial libraries and the derivation of QSAR models. Particular emphasis has been given to recent methods involving the use of multiobjective EAs. These are methods specifically designed for solving multiobjective optimisation problems in which different objectives are handled independently rather than by being combined into a single objective, which has been the more traditional approach. The MOGA and MOGP methods described here were originally developed for decision support in engineering control systems and have been found to be well suited to the current applications. It is clear that computer-aided drug design can benefit enormously from such sharing of ideas across different fields.

5 References

1. Muegge I, Rarey M (2001) Small molecule docking and scoring. In: Lipkowitz KB, Boyd DB (eds) Reviews in computational chemistry, vol. 17. VCH, New York
2. Gillet VJ, Johnson AP (1998) Structure generation for de novo design. In: Martin YC, Willett P (eds) Designing bioactive molecules. American Chemical Society, Washington, DC
3. Kubinyi H (2002) Quant Struct Act Relat 21:348
4. Bajorath J (2002) Nature Rev Drug Discov 1:882
5. Guner OF (ed) (2000) Pharmacophore perception, development, and use in drug design. International University Line, La Jolla, CA
6. Barnard JM, Downs GM, Willett P (1998) J Chem Inf Comput Sci 38:983
7. Ghose AK, Viswanadhan VN (eds) (2001) Combinatorial library design and evaluation. Principles, software tools and applications in drug discovery. Dekker, New York
8. Valler MJ, Green D (2000) Drug Discovery Today 5:286
9. Walters WP, Stahl MT, Murcko MA (1998) Drug Discovery Today 3:160
10. Bohm H-J, Schneider G (eds) (2000) Virtual screening for bioactive molecules. Wiley-VCH, Weinheim
11. Clark DE (ed) (2000) Evolutionary algorithms in molecular design. Wiley-VCH, Weinheim
12. Gillet VJ, Willett P, Bradshaw J (1997) J Chem Inf Comput Sci 37:731
13. Dunbar JB Jr (1997) Perspect Drug Discovery Design 7/8:51
14. Lajiness MS (1997) Perspect Drug Discovery Design 7/8:65
15. Mason JS, Pickett SD (1997) Perspect Drug Discovery Design 7/8:85
16. Brown RD (1997) Perspect Drug Discovery Design 7/8:31
17. Downs G M (2003) Molecular descriptors. In: Bultinck P, De Winter H, Langenaeker W, Tollenaere JP (eds) Computational medicinal chemistry and drug discovery. Dekker, New York
18. Leach AR, Gillet VJ (2003) An introduction to chemoinformatics. Luwer, Dordrecht
19. Brown RD, Clark DE (1998) Exp Opin Ther Patents 8:1447
20. Weber L (2000) Molecular diversity analysis and combinatorial library design In: Clark DE (ed) Evolutionary algorithms in molecular design. Wiley-VCH, Weinheim
21. Sheridan RP, SanFeliciano SG, Kearsley SK (2000) J Mol Graph Model 18:320
22. Gillet VJ, Nicolotti O (2000) Perspect Drug Discovery Design 20:265
23. Jamois EA, Hassan M, Waldman M (2000) J Chem Inf Comput Sci 40:63
24. Sheridan RP, Kearsley SK (1995) J Chem Inf Comput Sci 35:310
25. Zheng W, Cho SJ, Tropsha A (1998) J Chem Inf Comput Sci 38:251
26. Cho SJ, Zheng W, Tropsha A (1998) J Chem Inf Comput Sci 38:259
27. Weber L, Wallbaum S, Broger C, Gubernator K (1995) Angew Chem Int Ed Engl 107:2453
28. Singh J, Ator MA, Jaeger EP, Allen MP, Whipple DA, Soloweij JE, Chowdhary S, Treasurywala AM (1996) J Am Chem Soc 118:1669
29. Yokobayashi Y, Ikebukuro K, McNiven S, Karube I (1996) J Chem Soc Perkin Trans 1:2435
30. Gobbi A, Poppinger D (1998) Biotechnol Bioeng 61:47
31. Brown RD, Martin YC (1997) J Med Chem 40:2304
32. Gillet VJ, Willett P, Bradshaw J (1999) J Chem Inf Comput Sci 39:167
33. Lewis R, Pickett SD, Clark DE (2000) Computer-aided molecular diversity analysis and combinatorial library design. In: Lipkowitz KB, Boyd DB (eds) Reviews in computational chemistry, vol. 16. VCH, New York
34. Brown JD, Hassan M, Waldman M (2000) J Mol Graph Model 18:427
35. Zheng W, Hung ST, Saunders JT, Seibel GL (2000) PICCOLO: A tool for combinatorial library design via multicriterion optimization. In: Atlman RB, Dunkar AK, Hunter L, Lauderdale K, Klein TE (eds). Pacific symposium on biocomputing 2000. World Scientific, Singapore
36. Agrafiotis DK (2002) J Comput-Aided Mol Design 5/6:335
37. Fonseca CM, Fleming PJ (1995) An overview of evolutionary algorithms in multiobjective optimization. In: De Jong K (ed) Evolutionary computation, vol 3, no 1. The Massachusetts Institute of Technology

38. Fonseca CM, Fleming PJ (1993) Genetic algorithms for multiobjective optimization: formulation, discussion and generalisation. In: Forrest S (ed) Genetic algorithms: proceedings of the fifth international conference. Kaufmann, San Mateo, CA
39. Coello Coello CA (2000) ACM Computing Surveys 32:109
40. Gillet VJ, Khatib W, Willett P, Fleming PJ, Green DVS (2002) J Chem Inf Comput Sci 42:375
41. Gillet VJ, Willett P, Fleming P, Green DVS (2002) J Mol Graph Model 20:491
42. Wright T, GilletVJ, Green DVS, Pickett SD (2003) J Chem Inf Comput Sci 43:381
43. Hansch C, Fujita T (1964) J Am Chem Soc 86:1616
44. Kubinyi H (1997) Drug Discovery Today 2:457
45. So S-S (2000) Quantitative structure-activity relationships. In: Clark DE (ed) Evolutionary algorithms in molecular design. Wiley-VCH, Weinheim
46. Rogers DR, Hopfinger AJ (1994) J Chem Inf Comput Sci 34:854
47. Selwood DL, Livingstone DJ, Comley JC, O'Dowd BA, Hudson AT, Jackson P, Jandu KS, Rose VS, Stables JN (1990) J Med Chem 33:136
48. Kubinyi H (1994) Quant Struct Act Relat 13:285
49. Kubinyi H (1994) Quant Struct Act Relat 13:393
50. Luke BT (2000) J Mol Struct (THEOCHEM) 507:229
51. So S, Karplus M (1996) J Med Chem 39:1521
52. Nicolotti O, Gillet VJ, Fleming PJ, Green DVS (2002) J Med Chem 45:5069
53. Koza JR (1993) Genetic programming. MIT Press, Cambridge, MA
54. Rodríguez-Vázquez K, Fleming PJ (1998) Electron Lett 34:930

Structure and Bonding, Vol. 110 (2004): 153–175
DOI 10.1007/b13936

The Genetic Algorithm Approach to Protein Structure Prediction

Ron Unger

Faculty of Life Science, Bar-Ilan University, Ramat-Gan 52900, Israel
E-mail: ron@biocom1.ls.biu.ac.il

Abstract Predicting the three-dimensional structure of proteins from their linear sequence is one of the major challenges in modern biology. It is widely recognized that one of the major obstacles in addressing this question is that the "standard" computational approaches are not powerful enough to search for the correct structure in the huge conformational space. Genetic algorithms, a cooperative computational method, have been successful in many difficult computational tasks. Thus, it is not surprising that in recent years several studies were performed to explore the possibility of using genetic algorithms to address the protein structure prediction problem. In this review, a general framework of how genetic algorithms can be used for structure prediction is described. Using this framework, the significant studies that were published in recent years are discussed and compared. Applications of genetic algorithms to the related question of protein alignments are also mentioned. The rationale of why genetic algorithms are suitable for protein structure prediction is presented, and future improvements that are still needed are discussed.

Keywords Genetic algorithm · Protein structure prediction · Evolutionary algorithms · Alignment · Threading

1 **Introduction** 154

1.1 Genetic Algorithms 154
1.2 Protein Structure Prediction 157

2 **Genetic Algorithms for Protein Structure Prediction** 163

2.1 Representation 164
2.2 Genetic Operators 165
2.3 Fitness Function 165
2.4 Literature Examples 166

3 **Genetic Algorithms for Protein Alignments** 170

4 **Discussion** 172

5 **References** 173

Abbreviations

CASP	Critical assessment of methods of protein structure prediction
GA	Genetic algorithm
MC	Monte Carlo
MD	Molecular dynamics
rms	Root mean square

1 Introduction

Genetic algorithms (GAs) were initially introduced in the 1970s [1], and became popular in the late 1980s [2] for the solution of various hard computational problems. In a twist of scientific evolution, this computational method, which is based on evolutionary and biological principles, was reintroduced into the realm of biology and to structural biology problems in particular, in the 1990s. GAs have gained steady recognition as useful computational tools for addressing optimization tasks related to protein structures and in particular to protein structure prediction. In this review, we start with a short introduction to GAs and the terminology of this field. Next, we will describe the protein structure prediction problem and the traditional methods that have been employed for ab initio structure prediction. We will explain how GAs can be used to address this problem, and the advantages of the GA approach. Some examples of the use of GAs to predict protein structure will also be presented. Protein alignments will then be discussed, including aligning protein structures to each other, aligning protein sequences, and aligning structures with sequences (threading). (Docking of ligands to proteins, another related question is described elsewhere in this volume.) We will explain why we believe that GAs are especially suitable for these types of problems. Finally we will discuss what kind of improvements in applying GAs to protein structure prediction are most needed.

1.1 Genetic Algorithms

The GA approach is based on the observation that living systems adapt to their environment in an efficient manner. Thus, genetic processes involved in evolution actually perform a computational process of finding an optimal adaptation for a set of environmental conditions. Evolution works by using a large genetic pool of traits that are reproduced faithfully, but with some random variations that are subject to the process of natural selection. While there is no guarantee that the process will always find the optimal solution, it is evident that during the course of time it is powerful enough to select a combination of traits that enables the organism to function in its environment. The GA approach attempts to implement these fundamental ideas in other optimization problems. The principles of this approach were introduced by Holland in his seminal book *Adaptation in natural and artificial systems* [1]. The basic idea behind the GA search method is to maintain a population of solutions. This population is allowed to advance through successive generations in which the solutions are evolved via genetic operations. The size of the population is maintained by pruning in a manner that gives better survival and reproduction probabilities to more fit solutions, while maintaining large diversity within the population. This implies that the algorithm must utilize a fitness function that can express the quality of each solution as a numerical value. In many applications, possible solutions are represented as strings and are subject to three genetic operators: replication, crossover, and mutation. We will first present a specific, simple implementation of the method [2]. Many other versions have been suggested and analyzed, and we will discuss possible variations later.

The process starts with N random solutions encoded as strings of a fixed length at generation t_0; a fitness value is first calculated for each solution. For example, if the task is to find the shortest path in a graph, and each solution represents a different path, then the fitness value can be the length of that path. In the replication stage, N strings are replicated to form the next generation, t_1. The strings to be replicated are chosen (with repetitions!) from the current generation of solutions in proportion (usually linear) to their fitness, such that, for example, a solution that has a fitness value that is half the value of another solution will have half the chance of being selected for replication. Next comes the crossover stage: the new N strings are matched randomly in pairs (without repetitions) to obtain N/2 pairs. For each pair, a position along the string is randomly chosen as a cut point and the strings are swapped from that position onwards. This crossover process yields two new strings from the two old ones so that the number of strings is conserved. In addition, each string may be subject to mutation, which can change, at a predetermined rate, the individual values of its bits. This whole process constitutes the life cycle of one generation, and this life cycle (fitness evaluation, replication, crossover, and mutation) is repeated for many generations. The average performance of the population (as evaluated by the fitness function) will increase, until eventually some optimal or near-optimal solutions emerge. Thus, at the end of the search process, the population should contain a set of solutions with very good performance.

In this implementation, the bias towards solutions with better fit is achieved solely by imposing a greater chance to replicate for those solutions. This will present to the crossover stage an enhanced pool of solutions to "mix and match". The diversity of the population is maintained by the ability of the crossover operator to produce new solutions and by the ability of the mutation operator to modify existing solutions. As already mentioned, different versions of GAs differ in the specific way in which the solutions are represented, and the way the basic genetic operators are implemented. However, the two main principles remain: promoting better solutions while maintaining sufficient diversity within the population to facilitate the emergence of combinations of favorable features.

The crossover operation is the heart of the method. Technically, it is the simple exchange of parts of strings between pairs of solutions, but it has a large impact on the effectiveness of the search, since it allows exploration of regions of the search space not accessible to either of the two "parent" solutions. Through crossover operations, solutions can cooperate in the sense that favorable features from one solution can be mixed with others, where they can be further optimized. Cooperativity between solutions has been shown to have a very positive effect on the efficiency of search algorithms [3, 4].

While the basic computational framework is quite simple, there are many design and implementation details that might have a significant effect on the performance of the algorithm. Unfortunately, it seems that there are no general guidelines that might help the investigator match a given problem with a specific implementation. Thus, the choice of implementation is usually based on trial and error. In our experience, the most important factor determining the performance of the algorithm is how solutions are represented as objects that can be manipulated by the genetic operators. The original study by Holland used binary strings as a coding scheme, and bit manipulations as the genetic operators. This choice influenced many of the later implementations, although in principle there is no reason why more complex representations, ranging from vectors of real numbers to a more abstract data structure such as trees and graphs, could not be used. For the more complex representation, the genetic operators are more complicated than a flip of a binary bit or a "cut-and-paste" operation over strings. For example, if the representation is based on real numbers (rather than on a binary code), then a "mutation" might be implemented by a small random decrease or increase in the value of a number. It is true of course that real numbers can be represented by binary strings, and then be "mutated" by a bit operation, but this operation might change the value of the number to a variable degree depending on whether a more or less significant bit is affected. For example, in the example of finding the shortest path in a graph, a representation of a solution might be an ordered list of nodes along a given path. In this case a "mutation" operation might be a swap in the order of two nodes, and a crossover operation might be achieved by merging sublists from the lists that represent the parent solutions. It is difficult to predict a priori which representation is better, but it should be clear that in this example, as in many others, the difference in the representation can lead to a significant difference in performance. As already mentioned, the selection of the specific representation is usually empirical and based on trial end error.

One principle that does emerge from the work of Holland on strings (the schemata theorem) and from accumulated experience since is that it is important to place related features of the solution nearby in the representation and thus to reduce the chance that these features will be separated by a crossover event. This is of course true in biological evolution, where linked genes tend to be clustered along the chromosome. For example, consider the two alternative representations of a path in a graph. The first maintains the actual sequence of nodes along the path {3,1,6,2,5,4,7}, i.e. providing a direct description of the path, going from node number 3 to node number 1, then from node number 1 to node number 6, etc. The other alternative is to describe the path as an indexed list {2,4,1,6,5,3,7}, meaning that node number 1 is the second on the path, node number 2 is fourth on the path, node number 3 is the first on the path, etc. While the two representations contain exactly the same information, experience shows that the first representation is much more effective and enables faster discovery of the optimal solution. The reason is the locality aspect of the first representation, in which contiguous segments of the path remain contiguous in the representation, and thus are likely to remain associated during successive crossover operations. Thus, if a favorable segment is created, it is likely to be preserved during evolution. In the other representation, the notion of segment does not exist, and thus the search is much less efficient.

Another general issue to consider is the amount of external knowledge that is used by the algorithm. The "pure" approach requires that the only intervention applied will be granting a selective advantage to the fitter solutions such that they are more likely to participate in the genetic operations, while all other aspects of the process are left to random decisions. A more practical approach is to apply additional knowledge to guide and assist the algorithm. For example, crossover points might be chosen totally at random, but also could be biased towards preselected hotspots, based, for example, on success in previous generations or on external knowledge indicating that given positions are more suitable than others for crossovers.

Another major issue is how to prevent premature convergence at a local rather than at a global minimum. It is common that – during successive generations – one or very few solutions take over the population. Usually this happens much before the optimal solution is found, but once this happens the rate of evolution drops dramatically: crossover becomes meaningless and advances are achieved, if it all, at a very slow rate only by mutations. Several approaches have been suggested to avoid this situation. These include temporarily increasing the rate of mutations until the diversity of the populations is regained, isolating unrelated subpopulations and allowing them to interact with each other whenever a given subpopulation becomes frozen, and rejecting new solutions if they are too similar to solutions that already exist in the population.

In addition to these general policy decisions, there are several more technical decisions that must be made in implementing GAs. Among them is the trade-off, given limited computer resources, between the size of the population (i.e. the number of individuals in each generation) and the number of generations allocated for the algorithm. The mutation rate and relative frequency of mutations versus crossovers is another parameter that must be optimized.

1.2 Protein Structure Prediction

Predicting the three-dimensional structure of a protein from its linear sequence is one of the major challenges in molecular biology. A protein is composed of a linear chain of amino acids linked by peptide bonds and folded into a specific three dimensional structure. There are 20 amino acids which can be divided into several classes on the basis of size and other physical and chemical properties. The main classification is into hydrophobic residues, which interact poorly with the solvating water molecules, and hydrophilic residues, which have the ability to form hydrogen bonds with water. Each amino acid (or residue) consists of a common main-chain part, containing the atoms N, C, O, $C\alpha$ and two hydrogen atoms, and a specific side chain. The amino acids are joined through the peptide bond, the planar CO–NH group. The two dihedral angles, φ and ψ on each side of the $C\alpha$ atom, are the main degrees of freedom in forming the three dimensional trace

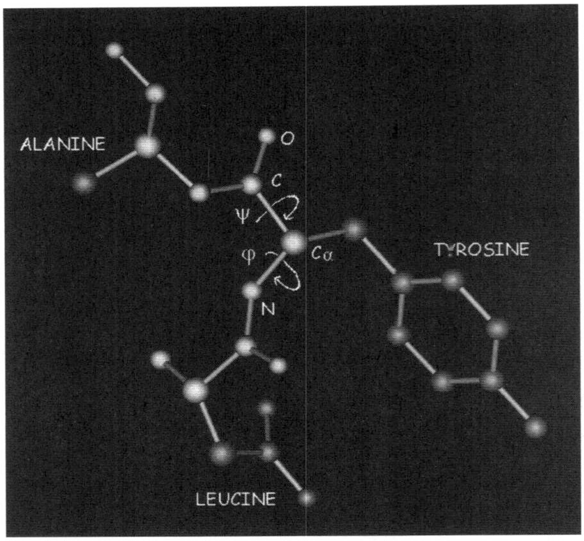

Fig. 1 A ball-and-stick model of a triplet of amino acids (valine, tyrosine, alanine) highlighting the geometry of the main chain (*light gray*). The main degrees of freedom of the main chain are the two rotatable dihedral angles φ, ψ around each Cα. The different side chains (*dark gray*) give each amino acid its specificity

of the polypeptide chain (Fig. 1). Owing to steric restrictions, these angles can have values only in specific domains in the φ, ψ space [5]. The side chains branch out of the main chain from the Cα atom and have additional degrees of freedom, called χ angles, which enable them to adjust their local conformation to their environment.

The cellular folding process starts while the nascent protein is synthesized on the ribosome, and often involves helper molecules known as chaperons. However, it was demonstrated by Anfinsen et al. [6] in a set of classical experiments that protein molecules are able to fold to their native structure in vitro without the presence of any additional molecules. Thus, the linear sequence of amino acids contains all the required information to achieve its unique three-dimensional structure (Fig. 2). The exquisite three-dimensional arrangement of proteins makes it clear that the folding is a process driven into low free-energy conformations where most of the amino acid can participate in favorable interaction according to their chemical nature, for example, packing of hydrophobic cores, matching salt bridges, and forming hydrogen bonds.

Anfinsen [7] proposed the "thermodynamic hypothesis", asserting that proteins fold to a conformation in which the free energy of the molecule is minimized. This hypothesis is commonly accepted and provides the basis for most of the methods for protein structure prediction.

Currently there are two methods to experimentally determine the three-dimensional structure (i.e. the three-dimensional coordinates of each atom) of a protein. The first method is X-ray crystallography. The protein must first be iso-

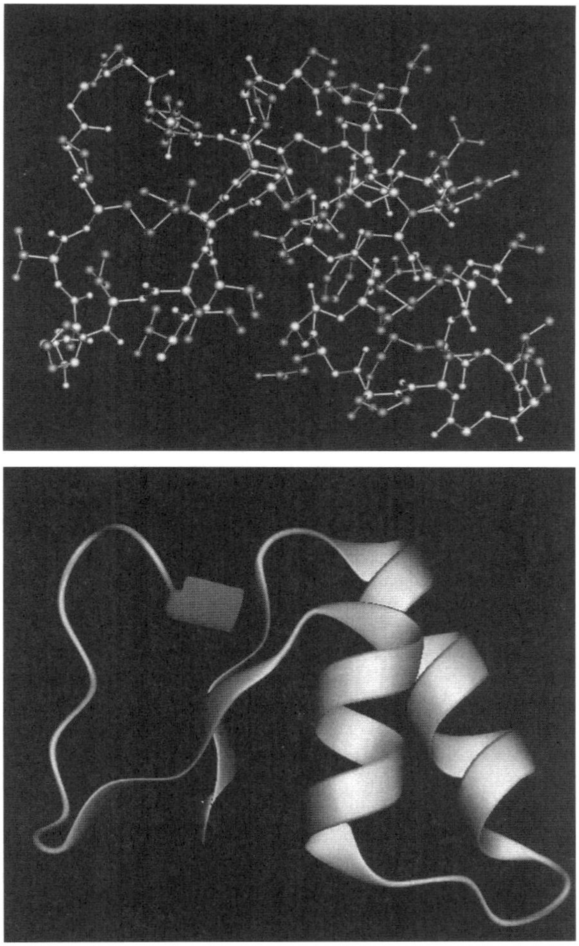

a

b

Fig. 2 **a** The detailed three-dimensional structure of crambin, a small (46-residue) plant seed protein (main chain in *light gray*, side chains in *darker gray*). **b** A cartoon view of the same protein. This view highlights the secondary structure decomposition of the structure with the two helices packing into each other along side a β-sheet

lated and highly purified. Then, a series of physical manipulations and a lot of patience are required to grow a crystal containing at least 10^{14} identical protein molecules ordered on a regular lattice. The crystal is then exposed to X-ray radiation and the diffraction pattern is recorded. From these reflections it is possible to deduce the actual three-dimensional electron density of the protein and thus to solve its structure. The second method is NMR, where the underlying principle is that by exciting one nucleus and measuring the coupling effect on a neighboring nucleus, one can estimate the distance between these nuclei. A series of such measured pairwise distances is used to reconstruct the full structure.

Many advances in these techniques have been suggested and employed in the last few years, mainly within the framework of structural genomics projects [8]. Nevertheless, since so many sequences of therapeutic or industrial interest are known, the gap between the number of known sequences and the number of known structures is widening. Thus, the need for a computational method enabling direct prediction of structure from sequence is greater than ever before.

In principle, the protein folding prediction problem can be solved in a very simple way. One could generate all the possible conformations a given protein might assume, compute the free energy for each conformation, and then pick the conformation with the lowest free energy as the "correct" native structure. This simple scheme has two major caveats.

First, the free energy of a given conformation cannot be calculated with sufficient accuracy. Various energy functions have been discussed and tested over the years, see, for example, Refs. [9, 10]; however, current energy functions are still not accurate enough. This can be demonstrated by two known, but often overlooked, facts. First, when native conformations of proteins from the protein data base whose three-dimensional structures were determined by high-resolution X ray measurements are subjected to energy minimization, their energy score tends to decrease dramatically by adjusting mainly local parameters such as bond length and bond angles, although the overall structure remains almost unchanged. This fact suggests that the current energy function equations overemphasize the minor details of the structure while giving insufficient weight to the more general features of the fold. It is also instructive to consider molecular dynamics (MD) simulations (see later) in which the starting point is the native conformation, but after nanoseconds of simulation time, the structure often drifts away from the native conformation, further indicating that the native conformation does not coincide with the conformation with minimal value of the current free-energy functions.

Second, and more relevant for our discussion here, no existing direct computational method is able to identify the conformation with the minimal free energy (regardless of the question whether the energy functions are accurate enough). The size of the conformational space is huge, i.e. exponential in the size of the protein. Even with a very modest estimation of three possible structural arrangements for each amino acid, the total number of conformations for a small protein of 100 amino acids is $3^{100}=10^{47}$, a number which is, and will remain for quite some time, far beyond the scanning capabilities of digital computers. Furthermore, it is not just the huge size of the search space that makes the problem difficult. There are other problems in which the search space is huge, yet efficient search algorithms can be employed. For example, while the number of paths in a graph is exponential (actually it scales as $N!$ for a graph with N nodes), there are simple, efficient algorithms with time complexity of N^3 to identify the shortest path in the graph [11]. Unfortunately, it was shown in several ways that the search problem embedded in protein folding determination belongs to the class of difficult optimization problems known as nondeterministic polynomial hard (NP-hard), for which no efficient polynomial algorithms are known or are likely to be discovered [12, 13].

Thus, it is clear that any search algorithm that attempts to address the protein folding problem must be considered as heuristics. Two search methods have traditionally been employed to address the protein folding problem: Molecular Dynamics (MD) and Monte Carlo (MC). These methods, especially MC, are described here in detail since, as we will see later, the GA approach incorporates many MC concepts. MD [14, 15] is a simulation method in which the protein system is placed in a random conformation and then the system reacts to forces that atoms exert on each other. The model assumes that as a result of these forces, atoms move in a Newtonian manner. Assuming that our description of the forces on the atomic level is accurate (which it is not, as noted earlier), following the trajectory of the system should lead to the native conformation. Besides the inaccuracies in the energy description there is one additional major caveat with this dynamic method: while one atom moves under the influence of all the other atoms in the system, the other atoms are also in motion; thus, the force fields through which a given atom is moving are constantly changing. The only way to reduce the effects of this problem is to recalculate the positions of each atom using a very short time slice (on the order of 10^{-14} s, which is on the same time scale as bond formation). The need to recalculate the forces in the system is the main bottleneck of the procedure. This calculation requires, in principle, N^2 calculations, where N is the number of atoms in the system, including both the atoms of the protein itself and the atoms of the water molecules that surround the protein and interact with it. For an average-sized protein with 150 amino acids, the number of atoms of the protein would be about 1,500, and the surrounding water molecules will add several thousand more. This constraint makes a simulation of the natural folding process, which takes about 1 s in nature, far beyond the reach of current computers. So far, simulations of only short intervals of the folding process, of the order of 10^{-8} s or 10^{-7} s are feasible [16].

While MD methods are based on the direct simulation of the natural folding process, MC algorithms [17, 18] are based on minimization of an energy function, through a path that does not necessarily follow the natural folding pathway. The minimization algorithm is based on taking a small conformational step and calculating the free energy of the new conformation. If the free energy is reduced compared to the old conformation (i.e. a downhill move), then the new conformation is accepted, and the search continues from there. If the free energy increases, (i.e. an uphill move) then a nondeterministic decision is made: the new conformation is accepted if (the Metropolis test [17])

$$\mathrm{md} < \exp\left[\frac{-(E_{\mathrm{new}} - E_{\mathrm{old}})}{kT}\right], \tag{1}$$

where rnd is a random number between 0 and 1, E_{old} and E_{new} are the free energies of the old and new conformation, respectively, T is the temperature, and k is Boltzmann's constant. In practice kT can be used as an arbitrary factor to control the fraction of uphill conformations that are accepted. If the new conformation is rejected, then the old conformation is retained and another random move is tested.

While MD methods almost by definition require a full atomic model of the protein and detailed energy function, MC methods can be used both on detailed models or on simplified models of proteins. These latter can range from a very abstract model in which chains that consist only of two types of amino acids are folded on a square 2D lattice [19] to almost realistic models in which proteins are represented by a fixed geometrical description of the main-chain atoms, and side chains are represented by a rotamer library [20]. The minimization takes place by manipulating the degrees of freedom of the system, namely, the dihedral angles of the main chain, and the rotamer selection of the side chain. These simplified representations are usually combined with a simplified energy function that describes the free energy of the system. Usually these energy functions represent mean force potentials based on statistics of frequencies of contacts between amino acids in a database of known structures [21]. For example, the relatively high frequency in known structures of arginine and aspartic acid pairs occurring a short distance apart relative to the random expectation indicates that such an interaction is favorable. The actual energy values are approximated by taking the logarithm of the normalized frequencies assuming that these frequencies reflect Bolzmann distributions of the energy of the contacts. As these, so-called empirical mean force, potentials are derived directly from the coordinates of known structures, they reflect all the free-energy components involved in protein folding, including van der Waals interactions, electrostatic forces, solvation energies, hydrophobic effects, and other entropic contributions. Because of their crude representation and their statistical nature, these potentials were shown not to be accurate enough to predict the native conformation. Thus, for known proteins, the native conformation does not coincide with the conformation represented by the lowest value of the potential. Yet, these potentials were shown to be useful in fold-recognition tasks, a topic which will be described later. In order to achieve more accurate mean force potentials, similar methods were used to derive the potential of interactions between functional groups rather than between complete amino acids [22]. It is still early to say whether these refined potentials will improve protein structure prediction.

What is a good prediction? The answer depends of course on the purpose of the prediction. Identifying the overall fold for understanding the function of a given protein requires less precision than designing an inhibitor for a given protein. The accuracy of the prediction (assuming of course that the real native structure is known for reference) is usually measured in terms of root-mean-square (rms) error, which measures the average distance between corresponding atoms after the prediction and the real structures have been superimposed on each other. In general, a prediction with rms deviations of about 6 Å is considered nonrandom, but not useful, rms deviations of 4–6 Å are considered meaningful, but not accurate, and rms deviations below 4 Å are considered good.

In recent years, the performance of prediction schemes has been evaluated at critical assessment of methods of protein structure prediction (CASP) meetings. CASP is a community-wide blind experiment in protein prediction [23]. In this test, the organizers collect sequences of proteins that are in the process of being experimentally solved, but whose structures are not yet known. These sequences are presented as a challenge to predictors, who must submit their structural predictions before the experimental structures become available. Previous CASP meetings have shown progress in the categories of homology modeling (where a very detailed structure of one protein is constructed on the basis of the known structure of similar proteins) and fold-recognition (where the task is to find on the basis of remote sequence similarity the general fold which the protein might assume). Minimal progress was achieved in the category of ab initio folding, predicting the structure for proteins for which there are no solved proteins with significant sequence similarity. However, in CASP4, which was held in 2000, a method based on the building-block approach, presented by Baker and his co-workers [24], was able to predict the structure of a small number of proteins with an rms below 4 Å. The prediction success was still rather poor and the method has significant limitations, yet it was the first demonstration of a successful systematic approach to protein structure prediction. For a recent general review of protein structure prediction methods see Ref. [25].

Progress in protein structure prediction is slow because both aspects of the problem, the energy function that must discriminate between the native structure and many decoys and the search algorithm to identify the conformation with the lowest energy, are fraught with difficulties. Furthermore, difficulties in each aspect reduce progress in the other. Until we have a search method that will enable us to identify the solutions with the lowest energy for a given energy function, we will not be able to determine whether the conformation with the minimal calculated energy coincides with the native conformation. On the other hand, until we develop an optimized energy function, we will not be able to verify that a particular search method is capable of finding the minimum of that specific function.

When discussing GAs for protein structure prediction, the same problem arises in making the distinction between evaluating the performance of the GA as a search tool and evaluating the performance of the associated energy function. Note that in almost all implementations, the energy function is also used as the fitness function of the GA, thus making the distinction between the energy function and the search algorithm even more difficult. At least for algorithmic design and analysis purposes, it is possible to detach the issues of the search from the issue of the energy function, by using a simple model where the optimal conformation is known by full enumeration of all conformations, or by tailoring the energy function to specifically prefer a given conformation (the Go model [26]).

2
Genetic Algorithms for Protein Structure Prediction

Using GAs to address the protein folding problem may be more effective than MC methods because they are less likely get caught in a local minimum: when folding a chain with a MC algorithm, which is based typically on changing a single amino acid, it is common to get into a situation where every single change is rejected because of a significant increase in free energy, and only a simultaneous change of several angles might enable further energy minimization. This kind of simultaneous change is provided naturally by the crossover operator of GA. In this section, we will first describe the general framework of how GA can be implemented to address protein structure prediction, and mention some of the decisions that must be made, which can influence the outcome. We will then describe some of the seminal studies in the field to illustrate both the strengths and limitations of this technique. Several good reviews on using GAs for protein structure prediction have been published in recent years [27–29].

2.1 Representation

The representation of solutions for GA implementation to address the protein structure prediction problem is surprisingly straightforward. As already mentioned, the polypeptide backbone of a protein has, to a large extent, a fixed geometry, and the main degrees of freedom in determining its three-dimensional conformation are the two dihedral angles φ and ψ on each side of the Cα atom. Thus, a protein can be represented as a set of pairs of values for these angles along the main chain $[(\varphi_1, \psi_1), (\varphi_2, \psi_2), (\varphi_3, \psi_3), \ldots, (\varphi_n, \psi_n)]$. This representation can be readily converted to regular Cartesian coordinates for the location of the Cα atoms. The dihedral angle representation of protein conformations can be used directly to describe possible "solutions" to the protein structure prediction problem. The process begins with a random set of conformations, which are allowed to evolve such that conformations with low energy values will be repeatedly selected and refined. Thus, with time, the quality of the population increases, many good potential structures are created, and hopefully the native structure will be among them. This representation maintains the advantages of locality of the representation, since local fragments of the structure are encoded in contiguous stretches. In some studies, the dihedral angles were stored and manipulated as real numbers. In other studies, the fact that dihedral angles occurring in proteins are restricted to a limited number of permitted values [5] enabled the choice of a panel of discrete dihedral angles [30], which could be encoded as integer values.

In lattice models, the location of each element on the lattice can be stored as a vector of coordinates $[(X_1, Y_1), (X_2, Y_2), (X_3, Y_3), \ldots, (X_n, Y_n)]$, where (X_i, Y_i) are the coordinates of element i on a two-dimensional lattice (a three-dimensional lattice will require three coordinates for each element). Since lattices enforce a fixed geometry on the conformations they contain, conformations can be encoded more efficiently by direction vectors leading from one atom (or element) to the next. For example in a two-dimensional square lattice, where every point has four neighbors, a conformation can be encoded simply by a set of numbers $(L_1, L_2, L_3, \ldots, L_n)$, where $L_i \in \{1, 2, 3, 4\}$ represents movement to the next point by going up, down, left, or right. Most applications of GAs to protein structure prediction utilize one of these representations.

These representations have one major drawback. They do not contain a mechanism that can ensure that the encoded structure is free of collisions, i.e. that the dihedral angles do not describe a trajectory that leads one atom to collide with another atom along the chain. Similarly, in a lattice, a representation based on direction vectors might describe walks that are not collision-free and could place atoms on already-occupied positions in the lattice. Thus, in most applications there is a need to include, in some form, an explicit procedure to detect collisions, and to decide how to address them. This is usually much more efficient to do on a lattice, where the embedding in the lattice permits a linear time algorithm to test for collisions simply by marking lattice points as free or occupied. A collision check is much more difficult to do with models that are not confined to lattices, where such a collision check has a square time complexity.

2.2 Genetic Operators

The genetic operator of replication is implemented by simply copying a solution from one generation to the next. The mutation operator introduces a change to the conformation. Thus, a simple way to introduce a mutation is to change the value of a single dihedral angle. Note, however, that this should be done with care, since even a small change in a dihedral value might have a large effect on the overall structure, since every dihedral angle is a hinge point around which the entire molecule is rotated. Furthermore, such a single change might cause collisions between many atoms since an entire arm of the structure is being rotated.

The crossover operation can be implemented simply by a "cut-and-paste" operation over the lists of the dihedral angles that represent the structure. In this way the "offspring" structure will contain part of each of its parents' structures. However, this is a very "risky" operation in the sense that it is likely to lead to conformations with internal collisions. Thus, almost every implementation needs to address this issue and come up with a way to control the problem. In many of the cases where the fused structure does not contain collisions, it is too open (i.e. not globular) and is not likely to be a good candidate for further modifications. To overcome these problems, many of the implementations include explicit quality control procedures that are applied to the structures produced in each new generation. Procedures could include exposing each generation of solutions to several rounds of standard energy minimization in an attempt to relieve collision, bad contacts, loose conformations, etc.

While these principles are shared by most studies, the composition of the different operators, and the manner and order in which they are applied, is – of course – different for each of the algorithms that have been developed, and give each one its special flavor.

2.3 Fitness Function

A wide variety of energy functions have been used as part of the various GA-based protein structure prediction protocols. These range from the hydrophobic potential in the simple HP lattice model [19] to energy models such as CHARMM, based on full fledged, detailed molecular mechanics [9]. Apparently, the ease by which various energy functions can be incorporated within the framework of GAs as fitness functions encouraged researchers to modify the energy function in very creative ways to include terms that are not used with the traditional methods for protein structure prediction.

2.4 Literature Examples

The first study to introduce GAs to the realm of protein structure prediction was that of Dandekar and Argos in 1992 [31]. The paper dealt with two subjects: the use of GAs to study protein sequence evolution, and the application of GAs to protein structure prediction. For protein structure prediction, a tetrahedral lattice was used, and structural information was encoded by direction vectors. The fitness function contained terms that encouraged strand formation and pairing and penalized steric clashes and nonglobular structures. It was shown that this procedure can form protein-like four-stranded bundles from generic sequences. In a subsequent refinement of this technique [32], an off-lattice simulation was described in which proteins were represented using bit strings that encoded discrete values of dihedral angles. Mutations were implemented by flipping bits in the encoding, resulting in switched regions in the dihedral angle space. Crossovers were achieved by random cut-and-paste operations over the representations. The fitness function used included both terms used in the original paper [31] and additional terms which tested agreement with experimental or predicted secondary structure assignment. The fitness function was optimized on a set of helical proteins with known structure. The results show a prediction within about 6 Å rms to the real structure for several small proteins. These results show prediction success which is better than random, but is still far from the precision considered accurate or useful. In Ref. [33], similar results were shown for modeling proteins which mainly include a β-sheet structure.

In a controversial study, Sun [34] was able to use a GA to achieve surprisingly good predictions for very small proteins, like melittin, with 26 residues, and for avian pancreatic polypeptide inhibitor, with 36 residues. The algorithm involved a very complicated scheme and was able to achieve accuracy of less than 2 Å versus the native conformation. However, careful analysis of this report suggests that the algorithm took advantage of the fact that the predicted proteins were actually included, in an indirect way, in the training phase that was used to parameterize the fitness function, and in a sense the GA procedure retrieved the known structure rather than predicted it.

Another set of early studies came from the work of Judson and coworkers [35, 36], which emphasized using GAs for search problems on small molecules and peptides, especially cyclic peptides. A dihedral angle representation was used for the peptides with values encoded as binary strings, and the energy function used the standard CHARMM force field. Mutations were implemented as bit flips and crossovers were introduced by a cut-and-paste of the strings. The small size of the system enabled a detailed investigation of the various parameters and policies chosen. In Ref. [37], a comparison between a GA and a direct search minimization was performed and showed the advantages and weaknesses of each method. As many concepts are shared between search problems on small peptides and complete proteins, these studies have contributed to subsequent attempts on full proteins.

We have studied [38] the use of GAs to fold proteins on a two-dimensional square lattice in the HP model [19] where proteins consist of only two types of paradigm "amino acids", hydrophobic and hydrophilic, and the energy function only rewards HH interactions by an energy score of –1. Clearly, in this model the optimal structure is one with the maximal number of HH interactions. For the GA, conformations were encoded as actual lattice coordinates, mutations were implemented by a rotation of the structure around a randomly selected coordinate, and crossover was implemented by choosing a pair of structures, and a random cutting point, and swapping paired structures at this cutting point. On a square lattice, there are three possible orientations by which the two fragments can be joined. All three possibilities were tested in order to find a valid, collision-free conformation. Another interesting quality control mechanism was introduced to the recombination process by requiring the fitness value of the offspring structure to be better, or at least not much worse, than the average fitness of its parents. This was implemented by performing a Metropolis test [17] (Eq. 1) comparing the energy of the daughter sequence to the averaged energy of its parents. If the structure was rejected, another pair of structures was selected and another fusion was attempted. This study enabled a systemic comparison of the performance of GA- versus MC-based approaches and demonstrated the superiority, at least in simple models, of GA over various implementations of MC. Further study [39] extended the results to a three-dimensional lattice. In Ref. [40] the effect of the frequency and quality of mutations was systematically tested. In most applications of GA to other problems, mutations are maintained at low rates. In our experiments using GA for protein structure determination, we found to our surprise that a higher rate of mutation is beneficial. It was further demonstrated that if quality control is applied to mutations such that each mutated conformation is subject to the Metropolis test and could be rejected, the performance improved even more. This gave rise to a notion that GA can be viewed as a cooperative parallel extension of the MC methodology. According to this concept, mutation can be considered as a single MC step, which is subject to quality control by the Metropolis test. Crossovers are considered as more complex changes in the state of the chain, which are followed by minimization steps to relieve clashes.

Bowie and Eisenberg [41] suggested a complicated scheme to predict the structure of small helical proteins in which GA search plays a pivotal role. The method starts by defining segments in the protein sequence in short, fixed-sized windows of nine residues, and also in larger, variable-sized windows of 15–25 residues. Each segment was then matched with structural fragments from the database with which the sequence is compatible, on the basis of their environment profile [42]. The pool of these structural fragments, encoded as strings of dihedral angles, was used as a source to build an initial population of structures. These structures were subject to a GA using the following procedure, Mutations were implemented as a small change in one dihedral angle. Crossovers were implemented by swapping the dihedral angles of the fragments between the parents. The fitness function used terms reflecting profile fit, accessible surface area, hydrophobicity, steric overlaps, and globularity. The terms were weighted in a way that would favor the native conformation as the conformation with the lowest energy. Under these conditions the method was able to predict the structure of several helical proteins with a deviation of as little as 2.5–4 Å from the correct structure.

As we have mentioned, most studies use dihedral angle representation of the protein and a cut-and-paste-type crossover operation. An interesting deviation was presented in the lattice model studied in Ref. [43]. Mutations were introduced as an MC step, where each move changed the local arrangement of short (2–8 residues) segments of the chain. The crossover operation was performed by selecting a random pair of parents and then creating an offspring through an averaging process: first the parents were superimposed on each other to ensure a common frame of reference and then the locations of corresponding atoms in each structure were averaged to produce an offspring that lay in the middle of its parents. Since the model is lattice-based, a refitting step was then required in order to place the structure of the offspring back within lattice coordinates. Since the emphasis in this study was on introducing and investigating this representation, the fitness function used was tailored specifically to ensure that the native structure would coincide with the minimum of the function. The method was compared to MC search and to standard GA, based on dihedral representation. For the examples presented in this study, it was shown that the Cartesian-space GA is more effective than standard GA implementations. The superiority of both GA methods over MC search was also demonstrated.

Another study, designed to evaluate a different variant of the crossover operator, was reported in Ref. [44]. A simple GA on a two-dimensional lattice model was used. The crossover operator coupled the best individuals, tested each possible crossover point, and chose the two best individuals for the next generation. It was shown that this "systematic crossover" was more efficient in identifying the global minimum than the standard random crossover protocol.

So far we have seen that GAs were shown in several controlled environments, for example, in simple lattice models or in cases where the energy function was tailored to guide the search to a specific target structure, to perform better than MC methods. The most serious effort to use GAs in a real prediction setting, although for short fragments within proteins, was presented by Moult and Pedersen. Their first goal [45] was to predict the structure of small fragments within proteins. These fragments were characterized as nucleation sites, or "early folding units" within proteins [46], i.e. fragments that are more likely to fold internally without influence from the rest of the structure. The full fragments (including side chains) were represented by their φ, ψ, and χ_i angles (χ_i determine the conformation of the side chains). The GA used only crossovers (no mutations were used) which included annealing of side-chain conformations at the crossover point to relieve collisions. The fitness function was based on point-charge electrostatics and exposed surface area which was parameterized using a database of known structures. The procedure produced good, low-energy conformations. For one of the fragments of length 22 amino acids, a close agreement with the experimental structure was reported. In a more comprehensive study [47], a similar algorithm was tested on a set of 28 peptide fragments, up to 14-residues long. The fragments were selected on the basis of experimental data and energetic criteria indicating their preference to adopt a nativelike structure independent of the presence of the rest of the protein. For 18 out of these 28 fragments, structure predictions with deviation less than 3 Å were achieved. In Ref. [48] the method was evaluated in the setting of the CASP2 meeting, as a blind test of protein structure predictions [23]. Twelve cases were simulated, including nine fragments and three complete proteins. The initial random population of solutions was biased to reflect the predicted secondary structure assignment for each sequence. Nevertheless, the prediction results, based on rms deviation from the real structure, were quite disappointing (in the range 6–11 Å). However, several of these predictions showed reasonable agreements for local structures but gross mistakes for the three-dimensional organization. This would suggest that the fitness function did not sufficiently consider long-range interactions.

In an intriguing paper [49], good prediction ability was claimed by a method in which supersecondary structural elements were predicted as suggested in Ref. [50], and then a GA-based method used them as constraints during the search for the native conformation. The protein was encoded by its φ, ψ, and χ_i angles, and the predicted supersecondary structural elements were confined to their predicted φ, ψ values. Crossovers were done by a cut-and-paste operation over the representation. There were two mutation operations available: one allowed a small change in the value of a single dihedral angle, and the other allowed complete random assignment of the dihedral angle values for a single amino acid. The fitness function was very simple and included terms for hydrophobic interactions and van der Waals contacts. This simple scheme was reported to achieve predicted accuracy ranging from 1.48 to 4.4 Å distance matrix error deviation from the native structure for five proteins of length 46–70 residues. Assuming, as the authors imply, that the distance matrix error (DME) measure is equivalent to the more commonly used rms error measure, then the results are surprisingly good. It is not clear what aspect of this scheme makes it so effective. Unfortunately no follow-up studies were conducted to validate these results.

Considering the generally poor ability of prediction methods, including those that are based on GAs, to provide accurate predictions based on sequence alone, the next studies [51–53] explored the possibility of including experimental data in the prediction scheme. In Ref. [51], distance constraints derived from NMR experiments were used to calculate the three-dimensional structure of proteins with the help of a GA for structure refinement. In this case, of course, the method is not a prediction scheme, but rather is used as a computational tool, like distance geometry algorithms, to identify a structure or structures which are compatible with the distance constraints.

In Ref. [52] it was demonstrated that experimentally derived structural information such as the existence of S-S bonds, protein side-chain ligands to iron-sulfur cages, cross-links between side chains, and conserved hydrophobic and catalytic residues, can be used by GAs to improve the quality of protein structure prediction. The improvement was significant, usually nudging the prediction closer to the target by more than 2 Å. However, even with this improvement, the overall prediction quality was still insufficient, usually off by more than 5 or 6 Å from the target structure. This was probably due to the small number and the diverse nature of the experimental constraints.

In Ref. [53], the coordination to zinc was used as the experimental constraint to guide the folding of several small zinc-finger domains. An elaborate scheme was used to define the secondary structure elements of the protein as a topology string, and then a GA was used to optimize this arrangement within the structural environment. The relative orientation of the secondary structure elements was calculated by a distance geometry algorithm. The fitness function consists of up to ten terms, including clash elimination, secondary structure packing, globularity, and zinc-binding coordination. A very interesting aspect of these energy terms is that the elements were normalized and then multiplied rather than added. This modification makes sure that all the terms have reasonable values, since even one bad term can deteriorate significantly the overall score.

3 Genetic Algorithms for Protein Alignments

Comparison of proteins may highlight regions in which the proteins are most similar. These conserved areas might represent the regions or domains of the proteins that are responsible for common function. Locating similarities between protein sequences is usually done using dynamic programming algorithms which are guaranteed to find the optimal alignment under a given set of costs for the sequence editing operation. The computational problem becomes more complicated when multiple (rather than pairwise) sequence alignments are needed.

Multiple sequence alignment was shown to be difficult [54]. Similarly, seeking structure alignment even between a pair of proteins, and clearly between multiple protein structures, is difficult. Another related difficult problem is threading: alignment of the sequence of one protein on the structure of another, which was also shown to be nondeterministic polynomial hard (NP-hard) [55]. Threading is useful for fold-recognition, a less ambitious task than ab initio folding, in which the goal is not to predict the detailed structure of the protein but rather to recognize its general fold, for example, by assignment of the protein to a known structural class. Because these are complex problems, it is not surprising that GAs have been used to address them. In these questions the representation issue is even more critical than in the protein structure prediction, where the dihedral angles set provides a "natural" solution.

SAGA [56] is a GA-based method for multiple sequence alignments. Multiple sequence alignments are represented as matrices in which each sequence occupies one row. The genetic operators (22 types of operators are used!) manipulate the insertions of gaps into the alignments. Since a multiple sequence alignment induces a pairwise alignment on each pair of sequences that participates in the alignment, then the fitness function simply sums the scores of the pairwise alignments. It was claimed that SAGA performs better than some of the common packages for multiple sequence alignment.

The issue of structure alignment was addressed in several studies. When two proteins with the same length and a very similar structure are compared, they can be aligned by a mathematical procedure [57] that finds the optimal rigid superposition between them. However, if the proteins differ in size or when their structures are only somewhat similar, then there is a need to consider introducing gaps in the alignment between them such that the regions where they are most similar could be aligned on each other (Fig. 3).

In Refs. [58, 59], a GA was used to produce a large number of initial rigid superpositions (using the six parameters of the superposition, three for rotation,

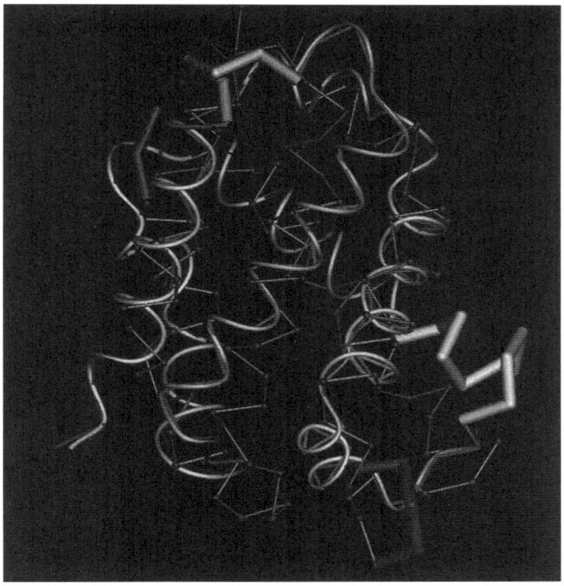

Fig. 3 Structural alignment of hemoglobin (β-chain) (the ribbon representation) with allophycocyanin (the ball-and-stick representation). The gaps in the structural alignment of one protein relative to the other are shown in a *thick line* representation. This alignment was calculated by the CE server (http://cl.sdsc.edu/ce.html)

and three for translation) as the manipulated objects. Then, a dynamic programming algorithm was used to find the best way to introduce gaps into the structural alignment. In Ref. [60], this method was extended to identify local structure similarities amongst a large number of structures. It was shown that the results are consistent with other methods of structural alignments.

In Ref. [61], structure alignment was addressed in a different way. Secondary structure elements were identified for each protein, and the structural alignment was done by matching, using a GA, these elements across the two structures. The representation was the paired list of secondary structure elements. The genetic operators changed the pairing of these elements to each other. A refinement stage was performed later to determine the exact boundaries of each secondary structure fragment. The results show very good agreement with high-quality alignments made by human experts based on careful structural examination.

In Refs. [62, 63] we studied the threading problem, the alignment of the sequence of one protein to the structure of another. Again the crux of the problem is where to introduce gaps in the alignment in one protein relative to the other. Threading was encoded as strings of numbers where 0 represents a deletion of a structural element relative to the sequence, 1 represents a match between the corresponding positions in the sequence and in the structure, and a number bigger than 1 represents insertion of one or more sequence residues relative to the structure. The genetic operators manipulated these strings by changing these numbers. The changes were done in a coordinated manner such that the string would always encode a valid alignment. In several test cases, it was shown that this method is capable of finding good alignments.

4 Discussion

GAs are efficient general search algorithms and as such are appropriate for any optimization problem, including problems related to protein folding. However, the superiority of GA over MC methods, which was demonstrated by many studies, suggests that the protein structure prediction problem is especially suited for the GA approach. This is quite intriguing since in reality protein folding occurs on the single-molecule level. Protein molecules fold individually (at least in vitro) as single molecules, and clearly not by a "mix-and-match" strategy on the population level.

The strength of the GA approach and its ability to describe many biological processes comes from its unique ability to model cooperative pathways. Protein folding is cooperative in many respects. First it is cooperative on the dynamic level, where semistable folded substructures on a single molecule come together to form the final structure. Protein folding is also "cooperative" on the interaction level, where molecular interactions including electrostatic, hydrophobic, van der Waals, etc., all contribute to the final structure. Furthermore, even with the current crude energy function models, the addition of a favorable interaction can usually be detected and rewarded, thus increasing the fitness of the structure that harbors this interaction. In time, this process will lead to the accumulation of conformations that include more and more favorable components.

If protein folding were a process in which many non-native interactions were first created, and then this "wrong" conformation were somehow transformed into the "correct" native structure, then GAs would probably fail. In other words, GAs work because they model processes that approach an optimum value in a continuous manner. In a set of experiments performed by Darby et al. [64], it was suggested that during folding of trypsin inhibitor, the "wrong" disulfide bridges must be formed first to achieve a non-native folding intermediate, and only then can the native structure emerge. This experiment was later repeated by other groups [65] but they failed to detect a significant accumulation of non-native conformations. The debate over the folding pathway of trypsin inhibitor is still active, but it seems that the requirement for disulfide formation makes this class of protein unique. In general models of folding (ranging from the diffusion/collision model [66] to folding funnels [67]), the common motif is gradual advancement of the molecules, along a folding path (in any way it is defined), and towards the final structure. This is compatible with an evolutionary algorithm for structure optimization. A protein may require two structural elements [x] and [y], as part of its correct conformation. The GA approach assumes that both [only x] and [only y] conformations still give a detectable advantage, though not as much as the conformation that has [x and y] together. This is consistent with the common view that a protein is folded through the creation of favorable local substructures that are assembled together to form the final functional protein, i.e.

these substructures can be considered as schemata [1] in the sequence that are consistently becoming more popular.

It is clear that GAs do not simulate the actual folding pathway of a single molecule; however, we may suggest the following view of GAs as being compatible with pathway behavior. We can refer to the many solutions in the GA system not as different molecules but as different conformations of the same molecule. In this framework a crossover operation may be interpreted as a decision of a single molecule, after "inspecting" many possible conformations for its C-terminal and N-terminal portions, on how to combine these two portions. Basically, each solution can be considered as a point on the folding pathway, while the genetic operators are used as vehicles to move between them.

As we have seen, many studies show that GAs are superior to MC and other search methods for protein structure prediction. However, no method based on GAs was able to demonstrate a significant ability to perform well in a real prediction setting. What kinds of improvements might be made to GA methods in order to improve their performance? One obvious aspect is improving the energy function. While this is a common problem for all prediction methods, an interesting possibility to explore within the GA framework is to make a distinction between the fitness function that is used to guide the production of the emerging solution and the energy function that is being used to select the final structure. In this way it might be possible to emphasize different aspects of the fitness function in different stages of folding.

Another possibility is to introduce explicit "memory" into the emerging substructure, such that substructures that have been advantageous to the structures that harbored them will get some level of immunity from changes. This can be achieved by biasing the selection of crossover points to respect the integrity of successful substructures or by making mutations less likely in these regions.

It seems as if the protein structure prediction problem is too difficult for a naïve "pure" implementation of GAs. The direction to go is to take advantage of the ability of the GA approach to incorporate various types of considerations when attacking this long-lasting problem.

Acknowledgements The help of Yair Horesh and Vered Unger in preparing this manuscript is highly appreciated.

5 References

1. Holland JH (1975) Adaptation in natural and artificial systems. The University of Michigan Press, Ann Harbor, MI
2. Goldberg DH (1985) Genetic algorithms in search, optimization and machine learning. Addison-Wesley, Reading, MA
3. Huberman BA (1990) Phys D 42:38
4. Clearwater SH, Huberman BA, Hogg T (1991) Science 254:1181
5. Ramakrishnan C, Ramachandran GN (1965) Biophys J 5:909
6. Anfinsen CB, Haber E, Sela M, White FH (1961) Proc Natl Acad Sci USA 47:1309
7. Anfinsen CB (1973) Science 181:223
8. Burley SK, Bonanno JB (2003) Methods Biochem Anal 44:591

9. Karplus M (1987) The prediction and analysis of mutant structures. In: Oxender DL, Fox CF (eds) Protein engineering. Liss, New York
10. Roterman IK, Lambert MH, Gibson KD, Scheraga HA (1989) J Biomol Struct Dyn 7:421
11. Even S (1979) Graph algorithms. Computer Science Press, Rockville, MD
12. Unger R, Moult J (1993) Bull Math Biol 55:1183
13. Berger B, Leighton TJ (1998) J Comput Biol 5:27
14. Levitt M (1982) Annu Rev Biophys Bioeng 11:251
15. Karplus M (2003) Biopolymers 68:350
16. Daggett V (2001) Methods Mol Biol 168:215
17. Metropolis N, Rosenbluth A, Rosenbluth M, Teller A, Teller E (1953) J Chem Phys 21:1087
18. Kirkpatrick S, Gellat CD, Vecchi MP (1983) Science 220:671
19. Dill KA (1990) Biochemistry 29:7133
20. Ponder JW, Richards FM (1987) J Mol Biol 193:775
21. Bryant SH, Lawrence CE (1993) Proteins 16:92
22. Samudrala R, Moult J (1998) J Mol Biol 6:895
23. Moult J, Pedersen JT, Judson R, Fidelis K (1995) Proteins 23:ii
24. Bonneau R, Tsai J, Ruczinski I, Chivian D, Rohl C, Strauss CE, Baker D (2001) Proteins Suppl 5:119
25. Baker D, Sali A (2001) Science 294:93
26. Go N, Taketomi H (1978) Proc Natl Acad Sci USA 75:559
27. Pedersen JT, Moult J (1996) Curr Opin Struct Biol 6:227
28. Le-Grand SM, Merz KM Jr (1994) The protein folding problem and tertiary structure prediction: the genetic algorithm and protein tertiary structure prediction. Birkhauser, Boston, p 109
29. Willett P (1995) Trends Biotechnol 13:516
30. Rooman MJ, Kocher JP, Wodak SJ (1991) J Mol Biol 5:961
31. Dandekar T, Argos P (1992) Protein Eng 5:637
32. Dandekar T, Argos P (1994) J Mol Biol 236:844
33. Dandekar T, Argos P (1996) J Mol Biol 1:645
34. Sun S (1993) Protein Sci 2:762
35. Judson RS, Jaeger EP, Treasurywala AM, Peterson ML (1993) J Comput Chem 14:1407
36. McGarrah DB, Judson RS (1993) J Comput Chem 14:1385
37. Meza JC, Judson RS, Faulkner TR, Treasurywala AM (1996) J Comput Chem 17:1142
38. Unger R, Moult J (1993) J Mol Biol 231:75
39. Unger R, Moult J (1993) Comput Aided Innovation New Mater 2:1283
40. Unger R, Moult J (1993) In: Proceedings of the 5th international conference on genetic algorithms (ICGA-93). Kaufmann, San Mateo, CA, p 581
41. Bowie JU, Eisenberg D (1994) Proc Natl Acad Sci USA 91:4436
42. Bowie JU, Luthy R, Eisenberg D (1991) Science 253:164
43. Rabow AA, Scheraga HA (1996) Protein Sci 5:1800
44. Konig R, Dandekar T (1999) Biosystems 50:17
45. Pedersen JT, Moult J (1995) Proteins 23:454
46. Unger R, Moult J (1991) Biochemistry 23:3816
47. Pedersen JT, Moult J (1997) J Mol Biol 269:240
48. Pedersen JT, Moult J (1997) Proteins 1:179
49. Cui Y, Chen RS, Wong WH (1998) Proteins 31:247
50. Sun S, Thomas PD, Dill KA (1995) Protein Eng 8:769
51. Bayley MJ, Jones G, Willett P, Williamson MP (1998) Protein Sci 7:491
52. Dandekar T, Argos P (1997) Protein Eng 10:877
53. Petersen K, Taylor WR (2003) J Mol Biol 325:1039
54. Just W (2001) J Comput Biol 8:615
55. Lathrop RH (1994) Protein Eng 7:1059
56. Notredame C, Holm L, Higgins DG (1998) Bioinformatics 14:407
57. Kabsch W (1976) Acta Crystallogr Sect B 32:922
58. May AC, Johnson MS (1994) Protein Eng 7:475

59. May AC, Johnson MS (1995) Protein Eng 8:873
60. Lehtonen JV, Denessiouk K, May AC, Johnson MS (1999) Proteins 34:341
61. Szustakowski JD, Weng Z (2000) Proteins 38:428
62. Yadgari J, Amir A, Unger R (1998) Proceedings of the international conference on intelligent systems for molecular biology, ISMB-98. AAAI, pp 193–202
63. Yadgari J, Amir A, Unger R (2001) J Constraints 6:271
64. Darby NJ, Morin PE, Talbo G, Creighton TE (1995) J Mol Biol 249:463
65. Weissman JS, Kim PS (1991) Science 253:1386
66. Karplus M, Weaver DL (1976) Nature 260:404
67. Onuchic JN, Wolynes PG, Luthey-Schulten Z, Socci ND (1995) Proc Natl Acad Sci USA 92:3626

Author Index Volumes 101–110

Alajarin M, see Turner DR (2004) *108*: 97–168
Aldinger F, see Seifert HJ (2002) *101*: 1–58
Aliev AE, Harris KDM (2004) Probing Hydrogen Bonding in Solids Using State NMR Spectroscopy *108*: 1–54
Alloul H, see Brouet V (2004) *109*: in preparation
Amstutz N, see Hauser A (2003) *106*: 81–96
Anitha S, Rao KSJ (2003) The Cemplexity of Aluminium-DANN Interactions: Relevance to Alheimer's and Other Neurological Diseases. *104*: 79–98
Aramburu JA, see Moreno M (2003) *106*: 127–152
Arcon D, Blinc R (2004) Jahn Teller Effect and Fullerene Ferromagnets *109*: in preparation
Atanasov M, Daul CA, Rauzy C (2003) A DFT Based Ligand Field Theory *106*: 97–125
Atwood DA, see Conley B (2003) *104*: 181–193
Atwood DA, Hutchison AR, Zhang Y (2003) Compounds Containing Five-Coordinate Group 13 Elements. *105*: 167–201

Barriuso MT, see Moreno M (2003) *106*: 127–152
Bellandi F, see Contreras RR (2003) *106*: 71–79
Berend K, van der Voet GB, de Wolff FA (2003) Acute Aluminium Intoxication. *104*: 1–58
Blinc R, see Arcon D (2004) *109*: in preparation
Bohrer D, see Schetinger MRC (2003) *104*: 99–138
Boulon G (2004) Optical Transitions of Trivalent Neodymium and Chromium Centers in $LiNbO_3$ Crystal Host Material *107*: in preparation
Bowlby BE, Di Bartolo B (2003) Spectroscopy of Trivalent Praseodymium in Barium Yttrium Fluoride *106*: 191–208
Brouet V, Alloul H, Gàràj S, Forrò L (2004) NMR Studies of Insulating, Metallic, and Superconducting Fullerides: Importance of Correlations and Jahn-Teller Distortions *109*: in preparation
Budzelaar PHM, Talarico G (2003) Insertion and β-Hydrogen Transfer at Aluminium. *105*: 141–165
Burrows AD (2004) Crystal Engineering Using Multiple Hydrogen Bonds *108*: 55–96

Cancines P, see Contreras RR (2003) *106*: 71–79
Cartwright HM (2004) An Introduction to Evolutionary Computation and Evolutionary Algorithms *110*: 1–32
Conley B, Atwood DA (2003) Fluoroaluminate Chemistry. *104*: 181–193
Contreras RR, Suárez T, Reyes M, Bellandi F, Cancines P, Moreno J, Shahgholi M, Di Bilio AJ, Gray HB, Fontal B (2003) Electronic Structures and Reduction Potentials of Cu(II) Complexes of [N,N′-alkyl-bis(ethyl-2-amino-1-cyclopentenecarbothioate)] (alkyl = ethyl, propyl, and butyl) *106*: 71–79

Daul CA, see Atanasov M (2003) *106*: 97–125
Day P (2003) Whereof Man Cannot Speak: Some Scientific Vocabulary of Michael Faraday and Klixbüll Jørgensen *106*: 7–18
Delahaye S, see Hauser A (2003) *106*: 81–96

Di Bartolo B, see Bowlby BE (2003) *106*: 191–208
Di Bilio AJ, see Contreras RR (2003) *106*: 71–79

Fontal B, see Contreras RR (2003) *106*: 71–79
Forrò L, see Brouet V (2004) *109*: in preparation
Frenking G, see Lein M (2003) *106*: 181–191
Frühauf S, see Roewer G (2002) *101*: 59–136
Frunzke J, see Lein M (2003) *106*: 181–191

Gàràj S, see Brouet V (2004) *109*: in preparation
Gillet VJ (2004) Applications of Evolutionary Computation in Drug Design *110*: 133–152
Golden MS, Pichler T, Rudolf P (2004) Charge Transfer and Bonding in Endohedral Fullerenes from High Energy Spectroscopy *109*: in preparation
Gray HB, see Contreras RR (2003) *106*: 71–79
Güdel HU, see Wenger OS (2003) *106*: 59–70
Gütlich P (2004) Recent Advances in Spin Crossover Research *107*: in preparation

Habershon S, see Harris KDM (2004) *110*: 55–94
Harris KDM, see Aliev (2004) *108*: 1–54
Harris KDM, Johnston RL, Habershon S (2004) Application of Evolutionary Computation in Structure Solution from Diffraction Data *110*: 55–94
Hartke B (2004) Application of Evolutionary Algorithm to Global Cluster Geometry Optimization *110*: 33–53
Haubner R, Wilhelm M, Weissenbacher R, Lux B (2002) Boron Nitrides – Properties, Synthesis and Applications. *102*: 1–46
Hauser A, Amstutz N, Delahaye S, Sadki A, Schenker S, Sieber R, Zerara M (2003) Fine Tuning the Electronic Properties of $[M(bpy)_3]^{2+}$ Complexes by Chemical Pressure ($M = Fe^{2+}$, Ru^{2+}, Co^{2+}, bpy = 2,2′-bipyridine) *106*: 81–96
Herrmann M, see Petzow G (2002) *102*: 47–166
Herzog U, see Roewer G (2002) *101*: 59–136
Hoggard PE (2003) Angular Overlap Model Parameters *106*: 37–57
Höpfl H (2002) Structure and Bonding in Boron Containing Macrocycles and Cages. *103*: 1–56
Hutchison AR, see Atwood DA (2003) *105*: 167–201

Iwasa Y, see Margadonna S (2004) *109*: in preparation

Jansen M, Jäschke B, Jäschke T (2002) Amorphous Multinary Ceramics in the Si-B-N-C System. *101*: 137–192
Jäschke B, see Jansen M (2002) *101*: 137–192
Jäschke T, see Jansen M (2002) *101*: 137–192
Jaworska M, Macyk W, Stasicka Z (2003) Structure, Spectroscopy and Photochemistry of the $[M(\eta^5\text{-}C_5H_5)(CO)_2]_2$ Complexes (M = Fe, Ru) *106*: 153–172
Johnston RL, see Harris KDM (2004) *110*: 55–94

Lein M, Frunzke J, Frenking G (2003) Christian Klixbüll Jørgensen and the Nature of the Chemical Bond in $HArF^t$ *106*: 181–191
Lever B (2004) Ruthenium Complexes of Non-Innocent Ligands; Aspects of Charge Transfer Spectroscopy *107*: in preparation
Linton DJ, Wheatley AEH (2003) The Synthesis and Structural Properties of Aluminium Oxide, Hydroxide and Organooxide Compounds. *105*: 67–139
Lux B, see Haubner R (2002) *102*: 1–46

Macyk W, see Jaworska M (2003) *106*: 153–172
Mahalakshmi L, Stalke D (2002) The R_2M^+ Group 13 Organometallic Fragment Chelated by P-centered Ligands. *103*: 85–116

Margadonna S, Iwasa Y, Takenobu T, Prassides K (2004) Structural and Electronic Properties of Selected Fulleride Salts *109*: in preparation
Miyake T, see Saito (2004) *109*: in preparation
Moreno J, see Contreras RR (2003) *106*: 71–79
Moreno M, Aramburu JA, Barriuso MT (2003) Electronic Properties and Bonding in Transition Metal Complexes: Influence of Pressure *106*: 127–152
Morita M (2004) Photoluminescence and Excitation Energy Transfer of Rare Earth Ions in Nanoporous Xerogel and Sol-gel SiO_2 Glasses *107*: in preparation
Morsch VM, see Schetinger MRC (2003) *104*: 99–138
Mossin S, Weihe H (2003) Average One-Center Two-Electron Exchange Integrals and Exchange Interactions *106*: 173–180
Müller E, see Roewer G (2002) *101*: 59–136

Nishibori E, see Takata M (2004) *109*: in preparation

Oshiro S (2003) A New Effect of Aluminium on Iron Metabolism in Mammalian Cells *104*: 59–78

Pastor A, see Turner DR (2004) *108*: 97–168
Patočka J, see Strunecká A (2003) *104*: 139–180
Petzow G, Hermann M (2002) Silicon Nitride Ceramics. *102*: 47–166
Pichler T, see Golden MS (2004) *109*: in preparation
Power P (2002) Multiple Bonding Between Heavier Group 13 Elements. *103*: 57–84
Prassides K, see Margadonna S (2004) *109*: in preparation
Prato M, see Tagmatarchis N (2004) *109*: in preparation

Rao KSJ, see Anitha S (2003) *104*: 79–98
Rauzy C, see Atanasov (2003) *106*: 97–125
Reber C (2004) Allowed and Forbidden d-d Bands in Octahedral Coordination Compounds: Intensity Borrowing and Interference Dips in Absorption Spectra *107*: in preparation
Reinen D (2004) The Angular Overlap Model and Vibronic Coupling in Treating s-p and d-s Mixing – a DFT Study *107*: in preparation
Reisfeld R (2003) Rare Earth Ions, Their Spectroscopy of Cryptates and Related Complexes in Glasses *106*: 209–237
Reyes M, see Contreras RR (2003) *106*: 71–79
Riesen H (2004) Progress in Hole-Burning Spectroscopy of Coordination Compounds *107*: in preparation
Roewer G, Herzog U, Trommer K, Müller E, Frühauf S (2002) Silicon Carbide – A Survey of Synthetic Approaches, Properties and Applications. *101*: 59–136
Rudolf P, see Golden MS (2004) *109*: in preparation

Sadki A, see Hauser A (2003) *106*: 81–96
Saito S, Umemoto K, Miyake T (2004) Electronic Structure and Energetics of Fullerites, Fullerides, and Fullerene Polymers *109*: in preparation
Sakata M, see Takata M (2004) *109*: in preparation
Schäffer CE (2003) Axel Christian Klixbüll Jørgensen (1931–2001) *106*: 1–5
Schäffer CE (2004) Ligand-Field Theory, Reformulated and Revived by Density Functional Theory *107*: in preparation
Schenker S, see Hauser A (2003) *106*: 81–96
Schetinger MRC, Morsch VM, Bohrer D (2003) Aluminium: Interaction with Nucleotides and Nucleotidases and Analytical Aspects of Determination. *104*: 99–138
Schmidtke HH (2003) The Variation of Slater-Condon Parameters F^k and Racah Parameters B and C with Chemical Bonding in Transition Group Complexes *106*: 19–35
Schubert DM (2003) Borates in Industrial Use. *105*: 1–40
Schulz S (2002) Synthesis, Structure and Reactivity of Group 13/15 Compounds Containing the Heavier Elements of Group 15, Sb and Bi *103*: 117–166

Seifert HJ, Aldinger F (2002) Phase Equilibria in the Si-B-C-N System. *101*: 1–58
Shahgholi M, see Contreras RR (2003) *106*: 71–79
Shinohara H, see Takata M (2004) *109*: in preparation
Sieber R, see Hauser A (2003) *106*: 81–96
Stalke D, see Mahalakshmi L (2002) *103*: 85–116
Stasicka Z, see Jaworska M (2003) *106*: 153–172
Steed JW, see Turner DR (2004) *108*: 97–168
Strunecká A, Patocka J (2003) Aluminofluoride Complexes in the Etiology of Alzheimer's Disease. *104*: 139–180
Suárez T, see Contreras RR (2003) *106*: 71–79
Sundqvist B (2004) Polymeric Fullerene Phases Formed Under Pressure *109*: in preparation

Tagmatarchis N, Prato M (2004) Organofullerene Materials *109*: in preparation
Takata M, Nishibori E, Sakata M, Shinohara H (2004) Charge Density Level Structures of Endohedral Metallofullerenes by MEM/Rietveld Method *109*: in preparation
Takenobu T, see Margadonna S (2004) *109*: in preparation
Talarico G, see Budzelaar PHM (2003) *105*: 141–165
Trommer K, see Roewer G (2002) *101*: 59–136
Turner DR, Pastor A, Alajarin M, Steed JW (2004) Molecular Containers: Design Approaches and Applications *108*: 97–168

Uhl W (2003) Aluminium and Gallium Hydrazides. *105*: 41–66
Umemoto K, see Saito S (2004) *109*: in preparation
Unger R (2004) The Genetic Algorithm Approach to Protein Structure Prediction *110*: 153–175

van der Voet GB, see Berend K (2003) *104*: 1–58

Weihe H, see Mossin S (2003) *106*: 173–180
Weissenbacher R, see Haubner R (2002) *102*: 1–46
Wenger OS, Güdel HU (2003) Influence of Crystal Field Parameters on Near-Infrared to Visible Photon Upconversion in Ti^{2+} and Ni^{2+} Doped Halide Lattices *106*: 59–70
Wheatley AEH, see Linton DJ (2003) *105*: 67–139
Wilhelm M, see Haubner R (2002) *102*: 1–46
de Wolff FA, see Berend K (2003) *104*: 1–58
Woodley SM (2004) Prediction of Crystal Structures Using Evolutionary Algorithms and Related Techniques *110*: 95–132

Zerara M, see Hauser A (2003) *106*: 81–96
Zhang Y, see Atwood DA (2003) *105*: 167–201

Subject Index

Adatom clusters 44
ADMET 134
Algorithms, evolutionary 33, 40, 57, 153
–, genetic 59, 95, 153, 154
Alignment 153
Allele 20
Amino acids 157
Annealing 99
–, simulated 46, 95, 138, 141, 148
Arity 28
Asymmetric unit 58
Atomic clusters 33

Basin of attraction 85
Basin-hopping 45, 46, 108
Beta-sheet/turns 76, 77
Binary representation 100–102, 125
– –, grey 128
Binary strings 19
Black box 31
Boltzmann distribution 162
Bond valence, model 108, 113, 120
Buckingham potential 107
Building blocks 48

Candidate structures 98
CASP 162, 168
CASTEP 96
Children 14, 25
Chromosome 18
Cluster aufbau strategies 48
Clustering 137
Clusters, atomic/molecular 33
–, passivated 43
–, pure/mixed 42
Code bloat 28
Colour wheel 6
Combinatorial synthesis 135
Combinatorial tiling theory 105
Competition 11
Compound selection, dissimilarity-based 137
Conjugate gradients 109, 111
Convergence, premature 157
Cost functions 97, 100–104
– –, expensive 49
Crossover 15, 23, 27, 100–104
–, multiple/single-point 66
– operator 155, 165
–, two-point 23
–, wrap-around 23
Cross-validation 144
Crystallography 59
–, biological 89
Cut-and-paste 14, 15

Deformation methods 47
Degree of freedom, correlated 45
De-novo design 134
Deterministic method 3
Differential evolution 62, 90
Diffraction, neutron 57
–, powder 60, 88
–, X-ray 56
Dissimilarity, compound selection 137
Distance least squares 127
Distribution plot, evolutionary 80, 82
Diversity 15
–, population 100–104, 112, 113, 129, 155
DNA 100
–, simulated 101
Docking 134
Drug design 4

EAGER 63, 87
Electron density 58
Elitism 24, 106, 112, 113
Emergent behaviour 11
Enantiomers 115
Encapsulation 31
Energies, minimum 39
Energy functions 160
Energy-lid method 122
Energy-minimization methods 97

Environment 12
ES, two-membered ES 25
–, two-numbered ES 25
Euclidean distance 137
Evolution 11
–, Darwinian/Lamarckian 69
–, differential 62, 90
– strategies 25
Evolutionary algorithms 57
Evolutionary distribution plot 80, 82
Evolutionary progress plot 71, 74, 81
Evolutionary trajectory plot 80, 83
Ewald summation 107, 123, 124
Exclusion zones 95, 124–128
Exhaustive search 5, 7
Experience 10
Exponential growth 17

Feature selection 145
Fingerprints 137
Fitness 12, 21, 29
– function 64, 65
–, normalised 22
Folding units, early 168
Fold-recognition 162, 163
Force potentials, mean 162
Free energy 160
Fullerene 40
Function set 28

GALOPED 140
Gaussian distribution 26
Genetic algorithms 18, 59, 153
– –, parallel 87
Genetic function approximation (GFA) 145
Genetic programming (GP) 28, 146, 147
Genotype 18, 25
Geometry, molecular 19
– optimization 33
Global minimum 62, 85, 157
Glutamic acid 70
Gold 43
Grid, nonlinear 125, 128
– points 98–104, 113, 125
– search 62
Guiding function 84
GULP 96

Heuristic search 5
High-throughput screening 135
Hill-climbing 3, 7, 24
Homology modeling 162
Hybrid function 83
Hybrid hypersurface 83

Hypersurface 62, 85
Image analysis 31
Indexing 60, 88
Interatomic potentials 96, 97, 115, 117, 126
Intermediate population 67

Jahn-Teller effect 108, 110
Junk DNA 31

Lattice energy, minimizing 96
Lattice models 162, 164, 166
Learning 6, 17
Least mean squares deviations 105
Lennard-Jones clusters 39
– – potential/interactions 107, 121–124
Lighthouse 12
Linear regression 144
Linkage learning 48
Local minima 85, 157
Local optimization 37, 45
Loewenstein rule 105

Main group clusters 40
Materials, disordered 90
–, nanocrystalline 91
Mating 64, 66
Maze 30
Mercury clusters 42
Metropolis Monte Carlo 99, 117, 118, 121
Migration 88
Minima, local 85
MOGA 142
MOGP 146, 148
Molecular dynamics (MD) 99, 160, 161
Monomer frequency analysis 138
Monte Carlo (MC) 62, 161, 172
– –, Metropolis 99, 117, 118, 121
MoQSAR 146–150
MoSELECT 142
Move class operators 97, 99, 104
Multidimensional strings 19
Multiple linear regression 144
Multiple sequence alignment 170
MUSEUM 146
Mutation 11, 14, 15, 24, 64, 100
–, dynamic/static 68
– operator 155, 165
– rate 17, 24

Nanotechnology 34
Natural selection 64, 69
Neural network 31
Neutron diffraction 57
Niches 49
NMR 159, 169

Nonlinear grid 125, 128
Normalised fitness 22

Oligopeptides 75
One fifth success rule 27
Optimal solution 3, 26
Optimization 4
–, local 37, 45
– methods, deterministic 47

Parallel computing 87
Parents 14, 25
Pareto parameter/ranking 106, 142
Parse tree 28
Partitioning 137
Path, shortest 160
Pattern recognition 31
Permutation 31
Pharmacophore keys 137
Phenotype 18, 25
Polyatomic units 129
Polymorphs 70
Polynomial hard problem, nondeterministic 35, 160
Population, initial 12, 20, 25, 29
–, intermediate 67
Population diversity 36
Population drift 16
Population size 17, 30
Potential energy 83
– – function 96, 97, 115, 117, 126
Powder diffraction 60, 88
Principal components analysis (PCA) 145
Product-based design 137, 138
Progress plot, evolutionary 71, 74, 81
Proteins 89
– alignments 170
– folding 46
– structure prediction 153

R factor 60
Random numbers 11
Random search 7
Random walk 17
Reactant-based design 136, 137
Reactor, chemical 24
Real-valued strings 19
Regression, linear 144
Replication 155, 165
Representation problem 36, 37
Representations 156, 164
Rietveld method/program 62, 71, 98, 113, 115
Root-mean-square 162
Roulette wheel 21, 24

SA, standard 46
Scaling, exponential 35, 46
Schema theory 103
Screening, high-throughput 135
Search 6
– space 4, 11
SELECT 140, 141
Selection 14, 21
– pressure 11
Shell model 108, 118
Silicon clusters 43
Similarity 137
Simulated annealing 10, 62, 138, 141, 148
Solids, crystalline 57
Space, direct 58, 60
– group 57
–, reciprocal 58
Space annealing, conformational 47
Stagnation 88
Standard deviation 26
Stochastic method 3
Stochastic remainder 22
Strategy parameter 26
String 18
– fridge 24
Structural fragment 62, 65
Structure factor 61
Structure refinement 60
Structure solution 60
Structures, minimum 39
Subpopulations 87
Supercells 117, 118, 128
Surface structure 91

T sites 105, 126, 127
TABU search 7, 47, 49, 139
Tanimoto coefficient 137
TB model 41
Terminal node 28
Thermodynamic hypothesis 158
Threading 170, 171
Ti_8C_{12} 41
Titanium metcars 41
Topological indices 137
Tournament 112
– selection 22, 24
Trajectory plot, evolutionary 80, 83
Transition, bond type 42
–, structural 41
Transparency 31
Trial structures, quality 97
Two-point crossover 23

Uniform grid 98
Unit cell 57

van der Waals interactions 122
Variables, evolutionary 17
Virtual screening 135

Water clusters 44, 45
Wrap-around operator 23
Wrapper 29

X-ray crystallography 158
X-ray diffraction patterns 56, 89, 96–98, 113–120

Z matrix 129

Printing: Saladruck, Berlin
Binding: Stein+Lehmann, Berlin